Hyper Relativity

Hiten Shelar

Published by IMAGINE SPACETIME (www.imaginespacetime.com)

Contents

Preface

The faster one moves in space, the slower one moves in time, and the faster one moves in time, the slower one moves in space. This is the basic theme of this book. It is a relation between space and time where time is nothing but a dimension like space. A tug of war of "c" between space and time is what makes the entity of spacetime exist. Where "c" is the speed of light. Thus, the speed of light is the speed of existence.

Wherever there is spacetime, there is relativity. The creation of the universe is the beginning of relativity. The universe is just a "relativity" in Hyperspace (Higher Dimensions). Thus, Hyper-relativity (Name of the book). With this theme, we understand the universe better. This book gives one a "Hyperspace Perspective". And one can see beyond the spacetime itself.

The deep nature of spacetime and Hyperspace and how spacetime is different from Hyperspace is explained in this book. Relativity here is not only a topic of time dilation, length contraction. It's beyond these topics. The topic of relativity is about every point of the fabric of spacetime that we call our universe.

About The Book

The theory of relativity is divided into two parts. The general and the special theory of relativity. The Special Theory of Relativity is about how frames of reference change when one nears the speed of light "c". The general theory of relativity is about how mass-energy density warps the spacetime and spacetime tells mass to move. This book transcends these topics, but. It talks about how the speed of light is the absolute speed of existence. How the topic of relativity is the topic of every single point of the spacetime universe. Covering these topics, this book also dives into various reference frames of relativity. It not only covers the relativity with a broader perspective, but also covers various other applications of relativity.

Two people, one from the normal universe and the other from pseudo pseudo-universe, trying to touch each other through gravity. They want to touch each other's fingers physically, but they are not able to. The only way they can communicate with each other is by light and gravity.

Where there is relativity, there is the universe!

One

The Spacetime

The use of the words space and time is generally seen when we are referring to some event. Or when we are planning for something. For example, you need to catch your train. Now the two questions one asks are ”on what platform” and “at what time”. One needs to do this to not miss his train. Now, this was an example of one event, and now when we look at the universe, we see that the universe itself is a fabric of events. Yes, just because we can merely see space and not time doesn’t mean that our universe is only space. The universe is spacetime, where every single point represents an event. In other words, the universe is the fabric of events. This fabric is what we call the “spacetime”. Spacetime because it’s a fabric of events.

Two

The Hyperspace

To know the higher dimensions it is necessary to understand spacetime and the things talked in this book. The first three dimensions is something that we are very familiar with. A zero dimension is an invisible point since it has no dimensions. Next, when one connects two of these points by a straight line, we see a rise of the first dimension, a line! A line is one dimension, it just has length and no breadth and height. Now connect two of these lines edge to edge with another two lines, and one gets some quadrilateral which is 2-dimensional. Now, again connect two of the quadrilaterals like these (let's say a square) and we get a cube. Which is a 3-dimensional object. So as we go higher in planes we see higher dimensions. And now let us go beyond 3 dimensions. Connect two 3 dimensional cubes in a higher plane. Try to imagine it! No one can do it fully! And what you get is a 4 dimensional hypercube, the "Tesseract".

Zero dimensions: Zero dimensions simply mean nothing. Since it has no dimensions. Also looked like a point (.)

One dimension: When two points connect each other, it forms a dimension. The image below represents one dimension (a line).

Two dimensions: Two lines connect each other in an extended plane (1+1) forming a 2 dimensional square. The image below represents two dimensions (a square, which is a side of a cube).

Three dimensions: Connect two 2-dimensional figures (square in this case) in extended dimension (2+1) and now one creates a cube. A cube is a side of a 4-dimensional cube called Tesseract.

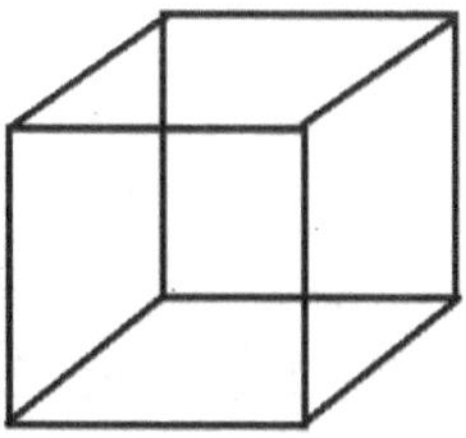

4 Dimensions: Imagine two cubes suspended in space in air side to side as shown in the image below. And what you

create is not 4-dimensions but a shadow of it. When 2 cubes are connected in an extended plane (3+1), a tesseract is formed. The image below shows the 3-dimensional projection of a 4-dimensional tesseract. Since the Human mind can't imagine more than 3 dimensions, we can't imagine tesseract as it is.

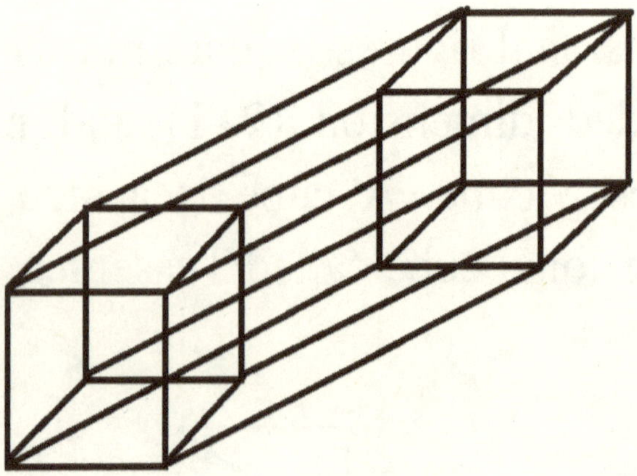

And there are infinite dimensions in the Hyperspace....

The human mind can't think of more than 3 dimensions. The world we see around us is

3 dimensional. Here everything has some length, breadth, and height. And above 3 dimensions is where the Hyperspace comes in. The Hyperspace means nothing but the infinite higher dimensions starting from four dimensions.

Let's get some physical notion of how hyperspace works through a story:

On a normal summer day, Joe comes back to New York after a long business work in Japan. This return time he decides to surprize his parents with a new bungalow in their home country US. The moment he was to surprize his parents with a new bungalow. The earthquake happens, The bungalow was cubical in shape with cubical shape rooms. In the earthquake, the cubical rooms begin to topple. And what lefts behind is the single cubical room. The other cubes just disappeared. When Joe cautiously enters the house to know what's happening in bungalow which just appears as a cube now, he gets surprized to see missing rooms in the undamaged state as he enters the house. But the house just appears as a single cube as seen from outside. How is this possible?

As Joe decided to check further, he climbs the stairs and finds the master bedroom above the entry way. Instead of finding the second floor, however he finds himself back on the ground floor. Thinking that the house is haunted, joe ran towards the main door. And instead of leading to the outside. The main door leads to another room. Joe gets faint. As he explore the house, he finds that each room is connected to an impossible series of other rooms. In the original bungalow, each room had a window to view the outside. And now all windows face other rooms and there is no outside.

Scared of what he witnessed, Joe slowly try all the doors of the house only to wind up in other rooms. Finally he studies, and decides to open the five venetian blinds and look outside. When he opens the first venetian blind, he finds himself peering at the Petronas twin tower. When opened the second venetian blind he was staring at the vast ocean. But upside down. Joe then opens the third venetian window he finds himself in Paris. Then he opened the fourth and he stares at the forest. Forest not on earth but on the earth like planet. After opening the fifth venetian blind, he was looking at the dessert from Mars. After haunting experience of the rooms of the house, Joe finally figures it all out. The earthquake had collapsed the joints of various cubical rooms and folded the bungalow in the four dimension.

From outside his bungalow looked as a single cube. The bungalow didn't collapse due to earthquake because, the joint between the cubes were rigid. And when the earthquake appeared, the rooms of the bungalow folded into a hypercube (Higher dimensional Shape). Joe got no way to figure out how to get out of the house. Until a similar earthquakes hits back and the house return back the normal with one room damaged. Joe again tries to open the main door of the house. And now he finds himself at the backyard of same house instead of front. Joe was happy, at least he could get out of the house.

This how it feels when one falls into the Hyperspace.

And similar happens when one tries to imagine “the time”. Because spacetime our universe is 4-dimensional. Something that can be considered as the Hyperspace structure. And something that is bizarre to imagine. The space part of the spacetime is that 3-dimensional part that we can actually see and perceive, and the “time” part is the 4-dimensional one. Although we don’t and can’t imagine the time. we know it exists due to its effects on the surroundings. Like the evolution of any system. So when start to treat our universe as spacetime, a hyperspace structure that is not able to be perceived as it is, then is when one really starts to know about how the universe works. Otherwise, the very questions we ask about our universe have no relevance.

So the spacetime universe is as bizarre as some Hyperspace structure. Let's keep that in mind.

Three

The Time

Time is a hyperspace dimension. Something that is not perceivable as it is. But there is something we can of course do to make it easy to deal with. Suppressing the dimensions is replacing one of the three dimensions (that we can think of) with the time dimension. So now instead of (x, y, z) coordinates, let's look at (x, y, *t*). Where *t* is the time coordinate of the time dimension. We can do this with any dimension of the infinite hyperspace. We do this so that we can bring that dimension under our domain of imagination and understanding.

Now to tell it exclusively, time is also a dimension like space. It's the same thing as space. One of the reasons we could replace it with our *z-coordinate.* So if we replace one of our 3 dimensions with time, how will we perceive the world? To understand this, we need to understand the difference in how we perceive the dimensions of space and time.

Like time we can't replace any quantity with the forth dimension. The time is chosen as the forth dimension as it shows it's relation with space when distorted. Suggesting us that time is nothing but the 4th dimension.

Four

How We Perceive Spacetime

Time is a dimension just like space, except that it's a hyperspace dimension and that we can't imagine it as the 4th dimension. But unlike other infinite dimensions of hyperspace, what makes the spacetime dimensions special is the way we behave in it. By "we" I mean every matter-energy in the universe. It is obvious that even if time is the same as space, we do know that it feels very different. I mean, when we talk about the three-dimensional space that we can imagine, we see it as the freedom of dimensions where we can actually roam and build the things, and do whatever we want. But when we think about time, we see it as the measure of change or the evolution of a system. So, how can the two of these be seen as the same thing?

Where there exists spacetime, there is a question of relativity. The concept of Spacetime, Relativity and matter-energy coexist. The Big bang was the beginning of these three things. With that the big bang, matter-energy and time came into existence instantaneously. Otherwise, the dimensions of hyperspace were always there. The very definition of relativity is the coexistence of spacetime and matter-energy. It's from where the topic starts. Further, the relativity becomes complex as the behavior of matter-energy in spacetime alters. This is the behavior of matter-energy in spactime.

Now let's understand the difference in our behavior in space and in time. Space and time are the same thing, the reason we see them differently has to do with a speed we travel within them. In the space we have control to slow down our speed when we travel, so we do have in time!

> **Hiten Shelar** @HitenShelar · 21m
> The very definition of matter-energy is it's tendency to travel at "C" (300,000 km/sec) either in time or in space or in both!
>
> #spacetime #relatvity

We all travel at a given speed within spacetime. There is no exception to this. Infact, the very definition of existence of matter-energy is its tendency to travel within spacetime at that given speed. The value of that speed is 299,792,458 m/s. This is called the speed of causality. So either one has to travel at this speed in space or in time. The one who stays at rest in familiar 3 spatial dimensions of spacetime do travel at this speed in time dimension. And that which rests in time, travels in 3 familiar spatial dimension.

The light travels at the same speed in space, so it doesn't travel in time at all. I mean time doesn't elapse for the light.

There are things in the universe which choose the value between these two extremes aswell. So it is possible to have a control on the time. I mean just move in space at some speed and then subtract it from 299,792,458 m/s and that would be the speed one travels in time. If one has to stop traveling in time, one has to travel at this same speed in space.

This relation of matter-energy with any set of 4 dimensions of infinite hyperspace gives the spacetime it's meaning.

This is theory of existence of relativity, spacetime and matter-energy. None of these three things comes into existence without creating the other.

Hiten Shelar
@HitenShelar

Relativity (4-dimensions of Hyperspace) = Spacetime = Existence Of Matter-energy.

They all co-exist!

Relativity here is the function of 4 dimensions of infinite Hyperspace.

The Beginning of relativity is the Beginning of the universe.

#relativity

Five

Matter-energy and time.

Well, what makes spacetime separate from the hyperspace is the behavior of matter in the random set of four dimensions in the Hyperspace. The speed of causality (299,792,458 m/s) also resembled “c” is the speed at which the matter-energy only spans over this selected set of four dimensions in the hyperspace.

So when we say time, we just mean that 4th dimension that we can’t perceive and have been moving through at “c” by being at rest in the space.

No matter-energy means nothing to move at "c", either in space or time. In such a case the concept of time is meaningless! The spacetime is no longer a spacetime, but just a random set of four dimensions from the infinite hyperspace.

Six

Moving In Time and Moving In Space

The universe is four-dimensional. So, it is a hyperspace structure that the human mind can't comprehend. The spacetime universe has four coordinates (x,y,z,*t*). Here *t* stands for time coordinate. We all move at c either in space or in time or in both simultaneously. Well, things become even crazier when we replace one of the 3 spatial dimensions with the time dimension. So let's say instead of (x,y,z,*t)* we have (x,y,*t*,z). So now the *z coordinate* switches with time and time switches with the *z coordinate*. What happens next is time becomes perceivable. One follows up on the motion he had in time wondering about why is he moving at the first place. Since a person was initially at rest in the z coordinate, and was traveling at “c” in the time dimension. As the coordinates got switched, he found himself continuing his speed “c”. And there is no wonder about that. As he was always doing that before the coordinates were switched!

Another interesting thing that happens is that he isn’t able to see anything that was ever part of his z-coordinate reference frame. Just like we can't think of hyperspace. Because the z coordinate has now become a 4th dimension, something that the human mind can’t think about. Check the following image.

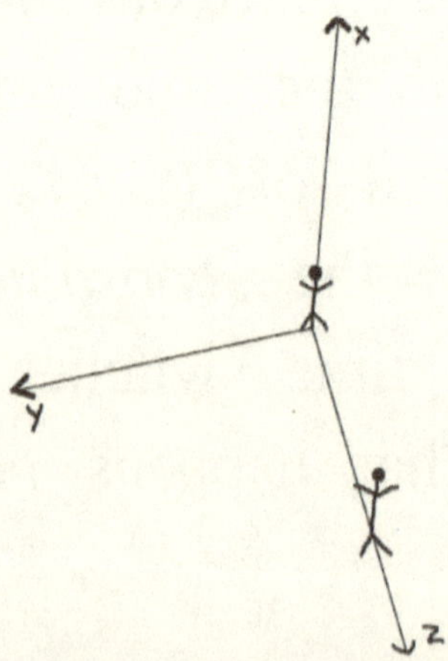

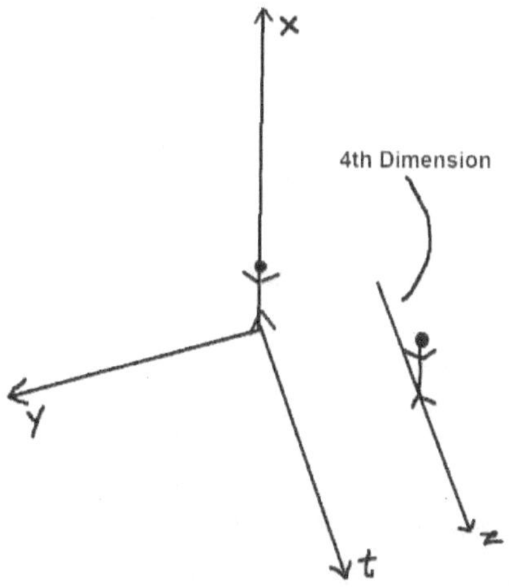

Two friends after coordinate systems are switched are unable to find each other.

That is for the new 4th dimension. But how would the new (x,y,*t)* world look like? Before we discuss that one must read the following.

The nature of time is highly debated. Some say that Past, Present and Future are already determined and while Quantum Physicists say that time is many-fingered, it fragments into multiple timelines in the multiple universes of the multiverse. While some say that the Past and Future don't exist and what exists is just the Present. In this book we are going to talk about all three of these.

Case no.1) So in this case, we assume that time is deterministic and everything is well set. When such a time dimension switches with the z coordinate, we do get to see in time! One can see his past, present, and future all at once in (x,y,*t)* world. So, one can see himself as a baby, a college student, and as an old man as one sees in space. And one finds himself traveling at "c" as previously discussed. One always has a choice to slow down in such a world as one does in (x,y,z,t) world. And besides this, everything would be as normal as (x,y,z,t) world, since time is just a dimension like space. But in this case, it is shown as encoding the past, present, and future.

Case no.2) The many-fingered nature of time says that time branches into the multiverse. And that the present that happens is chosen from one of these fragments of time. In such a universe a switch between the z and *t* axes produces nothing different than case 1 except that here the future will be presented in the form of what probably "can happen" instead of solid determinism. And the past as what happened before the present

Case no.3) That the present is the only real thing and that the past and future are just information. This concept is called Presentism. Where present can be the creator of the past "the creation under the cause of". This thing is well discussed in my book "The Next Level Observer Effect". Mainly it says that the reality itself means the "present" and past and future are just a piece of information.

In such a universe, if the z and *t* axes are switched, we would carry out our inherent momentum of time dimension. Same as case 1 and case 2. But this time, the time dimension won't display the past and future. It would just be treated as the dimension of space and nothing more. If one decelerates from "c" to rest in a new coordinate system, the z-coordinate becomes our new "time".

The common thing in all three cases discussed was the imperceivable of everything that lay in the observer's z coordinate.

This switch in dimensions is pure imagination. The imagination was included to teach the reader about “the time” as being same as “space”. And that it’s just matter-energy’s choice of behavior which give them different meanings.

This chapter also teaches how unfamiliar we are about the unthinkable dimension of our universe: the dimension of time.

Seven

How We Perceive Time

Time is a 4th hyperspace dimension. Something that humans can't imagine. As discussed before, matter-energy moves at "c" within spacetime. Some move at "c" in time while resting in space and some move at "c" in space while resting in time and while some travel fraction of "c" in space and other fraction in time.

We learnt that time is a 4th dimension and that we move at "c" in spacetime. But how can we relate it to the time we relate to the time on our watches? The 24hr time we talk about syncs with rotation of earth. When earth takes full rotation we call it "a day". And it takes 365 of these days for earth to revolve around the sun. And we call it "a year". So the time we refer on our clock is dialed with the motion of earth around the sun and its rotation on its own axis. This is parametrized time.

Every quantum particle is associated with the wave function, which has its own frequency. This frequency has units s^-1. This refers to the very fundamental clock of the universe. It is called the atomic clock. This is the example of "real time".

Now lets know diagrammatically, how the notion of time we generally refer.

In this concept we assume the deterministic nature of time.

See the image below

(*x,y,t*) World

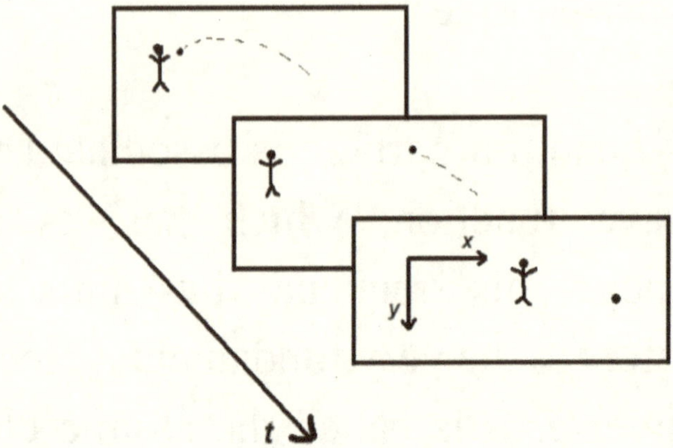

The image shows a boy throwing a ball in (x.y.t) world.

We chosen (x,y,*t*) world here for the sake of our imaginative capabilities. By this way we can get an analogue of what's happening in the higher dimensions.

As represented in the image, this is how we picture the dimension of time, “The evolution of things existing in space”. Here the time dimension represents the evolution of things existing in the x-y space.

The boy in the image is at rest in x-y plane and is traveling in time at “c”. But since he can’t imagine the time dimension being the 2-dimensional being, he is not aware of that.

One always moves in spacetime at “c”. Some do it in space, some in time and some fractionally in both. When one speeds up in space one slows down in time and when one speeds up in time one slows down in space.

Diagrammatically, this can be described as a boy moving in an x-y plane near to speed of causality “c” and having been slowing down in time. See the diagram below.

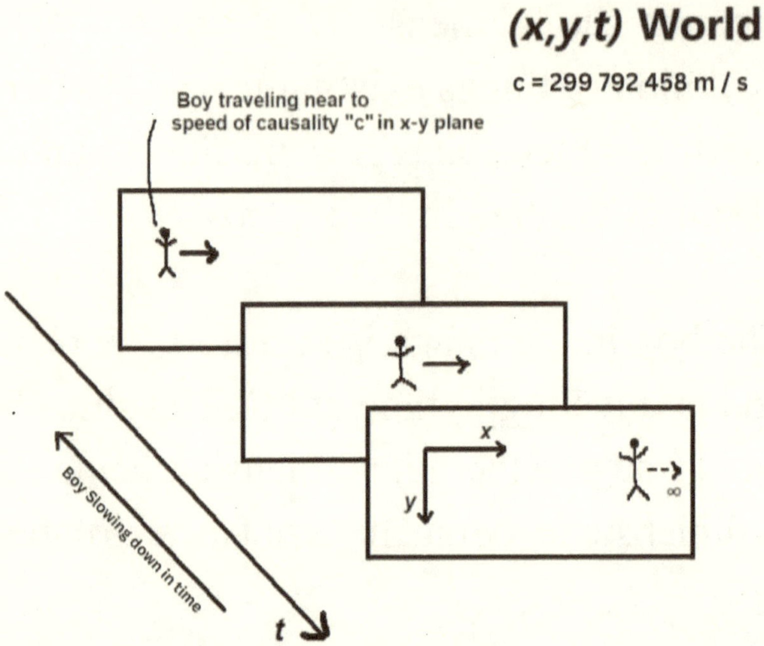

This slowing down in time is only significant when travels at speeds near to “c”.

In the examples above we considered (x,y,*t*) for the sake of our imagination. In reality, we have (x,y,z,*t*) plane. Where 4th dimension is out of the human imagination.

In the image, every plane is associated along with some value of time and this combination of a plane with some value of time is called an event.

Some theories say that spacetime can be divided into infinite planes like these. While quantum physics suggests that there is the smallest unit of division of time. It's called the Planck Time. It has a value of 5.4 × 10^-44 seconds. Which is an extremely tiny number. This was just for knowledge.

Well, this is how matter energy exists in spacetime, and that we perceive time as the fourth dimension.

Eight

When one speeds up in space and slows down in time

When one speeds up in space, one slows down in time, and when one speeds up in time, one slows down in space. This is the behavior of matter that underlines the very basic definition of its own existence. Not only this, but even time gets its meaning from this behavior of matter, as matter, energy, and spacetime coexist.

When we slow down in time, what difference does it make? How does the world appear in such a frame of reference? When one slows down in time, one doesn't notice any difference in his own frame of reference, and time elapses as it always was. This time is called the "Proper time". The proper time doesn't change in relativity. It's one's own time. But things become apparent when one compares this time with those we haven't slowed down in time.

The Person who has slowed down in time (by speeding up in space) when compares his time to the ones who are at rest in space witnesses a mismatch in time. So let's say there are two twin brothers, one of whom is an astronaut who sets up for a long 4-year journey in space at a speed near to "c". After a four-year journey when he returns to the earth and meets his twin brother who was at rest in space, he gets shocked. His twin brother is 7 years older than him! This was just an example of how slowing down in the time dimension feels like. The time has been dilated for one of the twins.

For the one who speeds up in space and slows down in time, the time for at rest appears to be shrunk. The idea will get more clear after checking the two diagrams below:

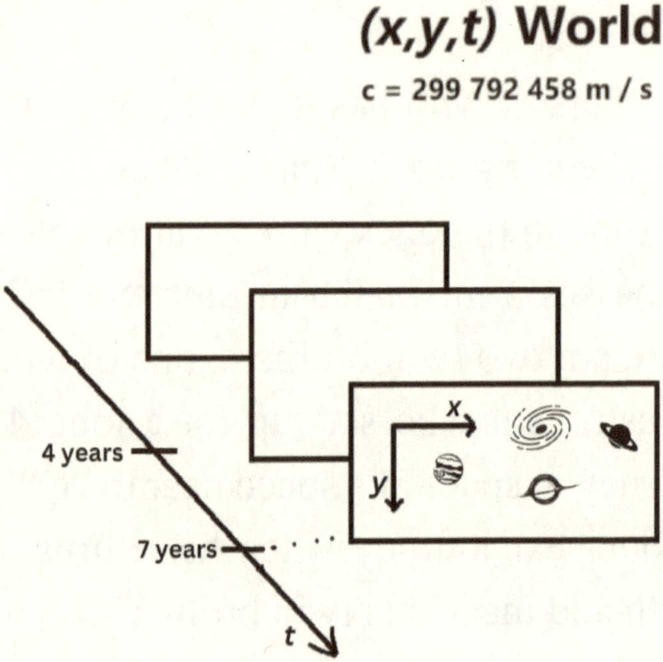

The image shows the universe on (x,y,t) coordinate system. This is how the universe appears to evolve for one of the two friends

(Alice) who is at rest in space. Evolution time for the state is 7 years in this frame of reference.

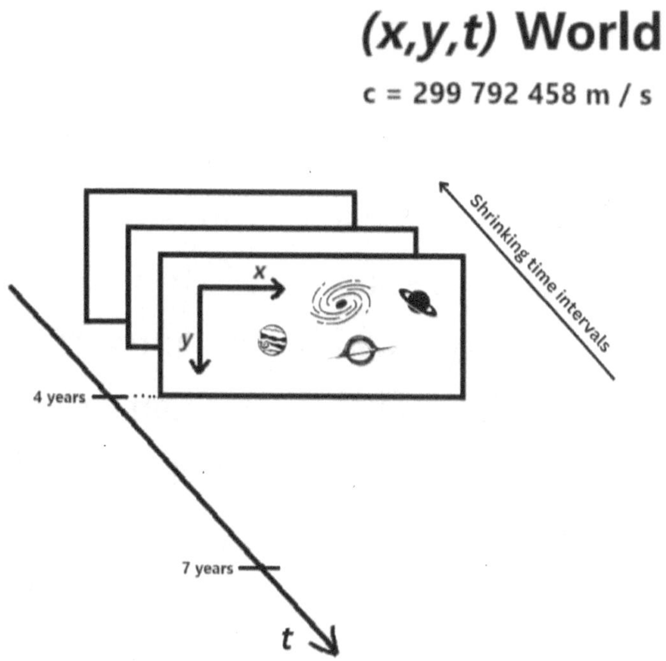

The image shows the shrinking of time intervals between adjacent x-y planes. This is how the universe is seen to evolve when the other friend (Bob) speeds up in space

traveling near "c". The Evolution time cuts from 7 years to 4 years in this frame.

When the two friends (Alice and Bob) meet, they don't agree on each other's observation of the time passage.

The Diagram shows the universe witnessed by two friends Alice and Bob. Alice decides to stay at rest in space, and Bob decides to take a long trip within the universe at a speed near "c". The First diagram shows how Alice sees the universe evolving around her and the other shows the case of Bob who was traveling at speed near "c".

When Bob was traveling close to the speed of causality "c", the universe was literally on fast-forward mode. This is the perspective or frame of reference of anyone who travels near to "c". When you speed up in space, you slow down in time. Since one has slowed down in time by speeding up in space near "c", the time for the universe around seems to runs fast. So, in the frame of reference of Bob, the time intervals between the adjacent x-y plane shrink along time as if it is a fast-forward.

When Alice and Bob met, it took 7 years in the frame of Alice and 4 years in the frame of Bob for the roundtrip. So at the end of the day, this makes Bob 3 years younger. This is the time dilation.

There is something more that happens when one travels at a speed near to "c". The length contraction! When one speeds up in space, he slows down in time. But along with that the traveller also notices a shrinking in length of things that are rest in space. This is called the length contraction. Why this happens? Let's learn it in the upcoming chapters.

Nine

The Speed of Causality

We all travel at the speed of "c" either in space or in time or fractionally in both. This speed is called the speed of causality. It's the maximum speed anything can travel. From the cause to effect, things can't travel faster than "c", thus the speed of causality. In other words, no information in the universe can be transmitted faster than this speed. The value of c is 299,792,458 m/s

When we speed up in space, we slow down in time and when we speed up in time we slow down in space. So wherever one speeds up, the maximum attainable speed is "c". When one travels at "c" in space, one stops moving in time completely. And for the one who travels in time at “c”, one stays at rest in space. The light or photon travels at “c” in space, so it is at rest in time. Time doesn’t elapse for light at all! Contrarily, the one that travels at “c” in time, stays at rest in space. And we know what staying at rest in space means.

The speed of causality “c” is one of the very basic fundamental constants of our universe. It’s one of the presumptions physics is based on. And it’s ok if physics doesn’t have a reason behind its exact value and not follow the treadmill of “whys” in physics.

The speed of causality is a quantity of “speed”. The speed talks about motion. Since matter-energy came into existence it’s been moving at “c” either in space or in time or fractionally in both. So, since the very Big Bang, every matter-energy is traveling at "c" within spacetime. The time of spacetime itself gets its meaning from the behavior of matter-energy that travels at "c" either in space or in time or fractionally both (As discussed in the beginning chapters of this book). Literally! Who thought the mere quantity of speed has to do so much with existence of the universe itself! The speed of causality is the fastest speed possible in the universe.

Ten

The Spacetime Dilation

When one speeds up in space, one slows down in time. And when one speeds up in time, one slows down in space. This transfer of the speed "c" from space to time or time to space is not smooth. As one reaches near to "c" the transfer of speed requires more and more energy. For anything traveling at 50% of "c" in space and 50% of c in time doesn't mean that the time will elapse at half of the rate of when one was at rest in space. The significant time dilation and other relativistic effects are only seen when one reaches closer to the speed of causality. This point is where it becomes not possible for anything to reach "c". As it then demands infinite energy. This is due to the transformational grid that is associated with speed transfer from space to time and time to space.

In layman's terms, space and time dilation is a byproduct of the transfer of “c” from space to time and time to space, but it’s not constant. The change is always associated with the transformation that shows the significant changes only when the speed transfer is near “c”. This transformation isn’t just mathematical, but is the impression of how geometry of spacetime continuum change when anything transfer speed from space to time or time to space. The length contraction is one of the results of this geometry transformation. When one travels near to speed of causality “c”, the lengths of the objects in the rest frame appears to be contracted. The length contraction is the byproduct of the same transformation.

There is one more way that can dilate spacetime. We know about it in this book ahead. Note that the spacetime dilation discussed here is because of transfer of “c” from space to time and time to space. We would learn why this happens in next few chapters.

Eleven

The Proper Time

The speed of causality is not a goal to chase. No amount of energy present in the universe can be focused on anything that has mass to reach this speed. Special relativity will never allow it. The speed of causality is the infinite speed of the universe. Even reaching 99.99999999 % of the speed of causality is just same as staying at rest in space.

Just like 99.99999999 % of infinity is infinitely far away from the infinity, Nothing that has mass, even at the speed this close to "c", is comparable to actually being at "c". The Proper time is a time that one always carries on their wristwatch. This time is how one feels time in one's own frame of reference. Even being at relativistic speeds and in extreme time dilations, this time on one's wristwatch always ticks. This time is called "The Proper Time". It exists for everything that travels at speeds below "c". At "c", the concept of proper time breaks! The particles, like photons, have no proper time. No particle that travels at "c" experiences the proper time.

The Higgs field became active 10^{-12} seconds after the Big Bang. In before that, no particle had mass. Time was just an untraveled road. Does this mean, the activation in Higgs field gave rise to the universe? The answer is no. The Big bang is the original creator. Lets see how.

Twelve

When Everything Travelled at "c"

Spacetime has no meaning without matter-energy and matter-energy cannot exist without spacetime. Thus, it's the tendency of matter-energy to speed up in space when it slows down in time and speed up in time when it slows down in space that defines the very existence of matter energy itself.

During or just after the Big Bang, the universe's temperature wasn't cold enough to allow the existence of particles that move slower than "c". Due to immense temperature, the Higgs field (discussed in this book) wasn't strong enough to give every particle its mass. So, every particle was massless like a photon. Traveling at “c”. They knew no time. Time was just another untraveled road for them.

This is like as if these particles don’t actually care if the spacetime universe exists. They had no charge and other properties that are associated with the four fundamental forces of the universe. They were closest to the definition of the purest form of energy (discussed ahead in detail in this book).

Only when many of these particles acquire mass the time by definition come into existence. After 10^{-12} seconds after the Big Bang, the Higgs field gave masses to the particles. Before that, the universe was almost the same as the Hyperspace. Not exactly the Hyperspace because the Big bang is the starting point of particles that kind of can exist without traveling in time, but needed the event of the creation of the universe to give them a picture of a quasi-particle. In this case, the quasi-particle is based on the system of our universe. In this case, this type of matter doesn't experience time, but its creation is based on the system called the universe, which is the Big Bang. This is true for any particle that travels at "c" in this case.

A quasi-particle is a particle that exists only in a given medium or system and can behave like real particles. The system here is our universe. Every particle was once this quasi-particle. After they got mass, they became secondary quasi-particles. Which is nothing but a quasi-particle whose existence is based on the world that's created from the existing quasi-particle. Thus, the existence of the universe marks a point in time even when there was nothing to travel in time. The quasi-particles even exist today. In the system of our universe, they travel at "c".

One might now ask what are the real particles then? The answer is Permanent Energy JACKET (discussed ahead in this book). After understanding the concept of permanent energy JACKET, one see the concept like “non-creation in creation”. At this point of this book, it’s not possible to understand it. Stay tuned as we dive into the book further.

Thirteen

The Mass

In the universe, things that have mass always travel below "c" and the things that have no mass always travel at "c". A particle with mass can travel at 99.9999999999999999999999% of "c" but never at "c"! Just after the Big bang, everything was massless, and so everything was traveling at "c". But as the universe cooled down, some particles got mass. And as we know, no positive mass can travel at "c" whatsoever. If the particles hadn't gotten mass after the universe cooled down, forming the atmos, stars, etc was impossible.

It is not the temperature of the universe that directly decides which particle would get what mass at what temperature. There is something called the Higgs field that decides this. The strength matter-energy interacting with the Higgs field is dependent on the Higgs constant. The value of the Higgs constant depends on the temperature of the universe. Just after the Big Bang, the temperature wasn't cool enough for the Higgs constant to give mass to the particles. But later, as the universe cooled down, the Higgs constant was strong enough to give a bunch of particles their "mass". Once the particle gets mass, it stops traveling at "c", since no mass can!

There are many other factors that influence the Higgs constant, the temperature is one of these. The Higgs field is not just theory, its existence was confirmed on 4 July 2024 at LHC when scientists could emerge and track the Higgs boson (a particle equivalent of the Higgs field).

Once the particle gets some mass it can't travel at "c". This "mass" that particles gets from the Higgs Field is called the "rest mass".

There is another type of mass that increases as one speeds up in space. As the particle with mass speeds up and gets closer to "c", its mass starts increasing. And one takes more and more energy to push the cart to "c". This happens until the mass becomes so massive that it takes infinite energy to attain "c". This increase in mass due to an increase in speed is called the relativistic mass.

Fourteen

Massless Particles and Particles with Mass.

The rest mass is a quantity that is totally assigned, by the Higgs field. The Higgs constant is not at the right value to give present photons some positive mass. The rest mass energy is all about the mass of the particle that have been received through the Higgs interaction. When we calculate the energy of a massless particle like a photon we take into account its frequency. We can't say the rest mass energy of a photon, as the photon has no rest mass.

Particles get their rest mass from the Higgs Field. Particles that interact weakly with the Higgs field tend to have smaller mass than particles that interact strongly with the Higgs Field.

The Higgs field gives fixed mass to the particles based on the Higgs constant and how strongly the particles interact with the Higgs field. This mass of the particle is called the rest mass. It never changes until the value of the Higgs constant itself changes. Which is a rare event even at the cosmic time scale. The rest mass energy is the measure of the rest mass of the particle. In the case of massless particles like photons, we calculate the energy content by its frequency.

The mechanisms that give particles the value of mass have to do a lot with force fields and symmetries associated with these force fields. Some particles don't gain mass by Higgs field due to their strong symmetries.

For example, the U1 gauge symmetry associated with electromagnetism doesn't allow photons to gain mass by the Higgs field. So it's a complex study that is out of the scope of this book.

The thing we need to know is that the Higgs field gives very basic rest mass to the particles. When the particles get the rest mass, they can't travel at "c" in space. When the particle speeds up closer to the speed of causality "c", it starts to gain extra mass called the relativistic mass. The relativistic mass is not due to the increase in matter content. One must keep this in mind.

The source of the relativistic mass is the same energy that one puts to accelerate anything closer to "c" as one gets closer to "c" (Energy mass equivalence discussed ahead in this book). On the other hand if one tries to accelerate a photon beyond "c", the energy gets accumulated in the form of an increase in frequency of the photon. Which increases the energy of the photon. The increase of mass (relativistic mass) and energy of the photon both are relative here. Since one is always at rest in one's own frame, the gain of relative mass is felt by anyone traveling closer to "c". He will notice it only when he tries to reach "c" and is never does so due to increased mass (relative mass).

Various particles are various manifestations of the 4-force fields of the universe. Each of these fields have some symmetries associated with them. The higgs field gives masses to the particles only when the force fields and their associated symmetries allow so with the right value of Higgs constant.

When one speeds up in space one slows down in time and when one slows down in time, one speeds up in space. This transfer of "c" is not as easy in general. The particles that gain some mass always travel between 0 to "c". While the massless particles always keep it up at "c" and never below that.

Anyways, this allows some of these particles to form atoms and structures like the sun and earth and others to transmit the energy of these stars to the earth.

Fifteen

Relativity

There are two types of perspectives, in the theory of relativity they are known as two frames of reference. One is inertial and the other, non-inertial. The inertial frame of reference means the rest frame of reference and the other, is the moving frame of reference. If the moving frame of reference is moving close to the speed of causality, the events appear to be relative in the universe. With the mentioned difference in the frames of reference, any given event appears differently for these two frames.

For example, a person at rest watches a man in a moving car throwing a ball outside in the direction the car is moving. The car is traveling at velocity "v", and the ball is thrown at velocity "b". Now according to the person at rest frame, the ball is traveling at v+b since it was thrown out of a moving car. While in the frame of a car driver, the speed of the ball comes out to be “b” and not “v+b”. The ball appears to travel faster in the frame of reference of a man who is at rest (Basic Principle of addition of velocities). Now replace the ball with a light laser. Now one might say that just like a ball, the light must be travelling at a faster speed in the frame of reference of the person at rest. But that’s not how the universe works! For the “speed” light is traveling at, there is something special! The speed of light doesn’t change unlike the ball thrown out of the running car. The speed of the car doesn’t add up in this case! The reason behind this is explained by the theory of relativity.

Time dilation, length contraction, increase in mass at relativistic speeds, etc which we discussed before in this book are explained by the theory of relativity. The theory of relativity is a theory of change in spacetime continuum. At a local scale it's referred as the special theory of relativity and at a larger scale, it's referred to as the general theory of relativity. Special of the "special relativity" in the sense that its effects are limited to the two frames we are comparing each other to. And "general" of general relativity in the sense that it applies to everything in the universe in the inertial or non-inertial frames of reference. The theory of relativity is the theory of events that are dependent on the frame of references. The theory joins a gap between two different measurements of the same event.

Hiten Shelar @HitenShelar · 04 Feb

Where there is relativity, there is the universe!

#physics #relativity

(Post on X.com) All of these will make sense ahead in this book

Let's now first complete the ongoing topic of the special theory of relativity.

Sixteen

The Transformations

We have two reference frames, one is inertial and the other non-inertial. An inertial frame is referred to as a rest frame. In an inertial frame, the object may appear to be at rest, but in real spacetime, the object is always moving in time. The object in the non-inertial frame when speeds up in space, it automatically slows down in time.

Talking about these two frames of reference, there are always disagreements in the measurement of any given particular event. Now the question that arises is which frame of reference is actually correct or real? The answer is no frame of reference is more real than another. In the theory of relativity, both inertial and non-inertial frames of reference are considered equally true and significant. It's just based on what coordinate system or coordinate grid we are using to analyze the two reference frames. These coordinate grids are imaginary and are only used to study motion. And they can be transformed into various shapes. For example, in our example, we could have assumed the man in the moving car in the inertial frame of reference and the one at rest to be in the moving frame in case if we transformed the coordinate grid the right way. The labels of the inertial frame and the non-inertial are decided by the coordinate grid we use. The coordinate grid can be transformed from one to another to get a completely different notion of the motion. But one exception to this is the speed of causality. Which travels at the same speed, no matter what reference frame we choose. The light travels at the speed of causality.

Even if one shines a beam of light from a car moving at 70% of “c” (let’s say), the speed of the car doesn’t add up. And the light will still be traveling at “c”.

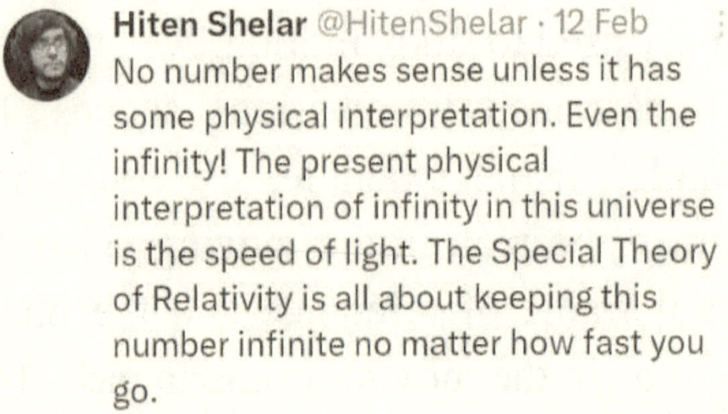
Hiten Shelar @HitenShelar · 12 Feb

No number makes sense unless it has some physical interpretation. Even the infinity! The present physical interpretation of infinity in this universe is the speed of light. The Special Theory of Relativity is all about keeping this number infinite no matter how fast you go.

The speed of causality is the infinite speed of spacetime. So anything that adds to infinity is still the infinity! The speed we can attain in this universe is 0 to infinity, where “c” is another name of infinity. This is how motion works in this universe.

There is a coordinate system that governs how the motion of the universe itself works. In special relativity, this coordinate system is the Minkowski spacetime. With some assumptions in this coordinate system, one can actually see how this coordinate system describes about how the motion works in this universe.

Ideally for comparing two reference frames that are non-relativistic (not close to "c"), we generally use the flat spacetime coordinate to describe the motion. But in the case of comparing an inertial and relativistic (close to "c") moving frame, we have to consider the Minkowski spacetime.

A straight line on a paper doesn't appear straight when the paper is wrapped up on a ball. This is an example of a transformation. Disturbing any coordinate grid in any arbitrary way can also be considered as the transformation. A rotation of a coordinate grid is also a transformation.

Unlike normal spacetime, the change in frames of reference in Minkowski spacetime is looked as a Hyperbolic rotation around the center of the Hyperbola. Where asymptotes of hyperbola represent the speed of causality. The transformation from one frame to the other in the Minkowski Spacetime is called the Lorentz transformation. Which are analogous to the Hyperbolic rotation.

Unlike circular rotations, where the full rotation costs 360 degrees, the hyperbolic rotations go indefinitely large. The angle in this hyperbolic rotation is called the rapidity. When the car goes from an inertial to a non-inertial frame, its rapidity increases in this coordinate system. The speed of light resembles the asymptotes of the hyperbola.

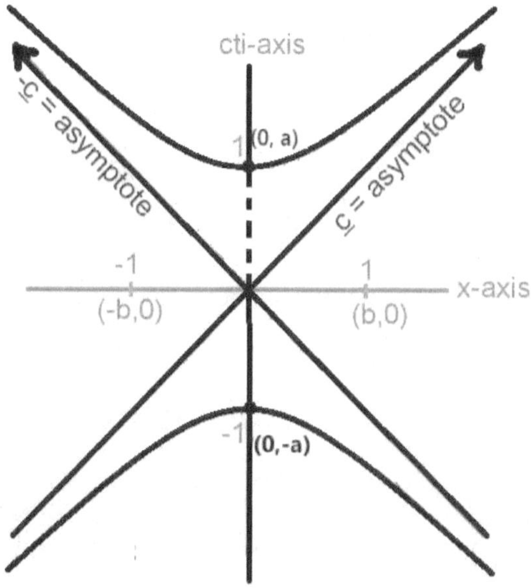

The hyperbola meets the asymptotes at infinity. So it takes infinite rapidity for anything to travel at "c". So the spacetime where the change in frame accounts for the increase in rapidity, it takes infinite rapidity to attain the speed of causality. This is why no mass in the universe can attain this speed. The speed of causality thus represents the infinite speed of the universe. And since the infinity is unchanged by addition, no speed adds to the speed of light that is traveling at the speed of causality.

The speed of causality is the asymptote in the Minkowski spacetime. Thus, light always travels at 45 degrees in this spacetime. This 45-degree angle marks a boundary on how curved the trajectory of a particle with mass can become in Minkowski spacetime. To note, which has to be lower than 45 degrees for anything that has mass. Adding one more dimension to the two-dimensional Minkowski diagram in the discussion. We get something called the "lightcone".

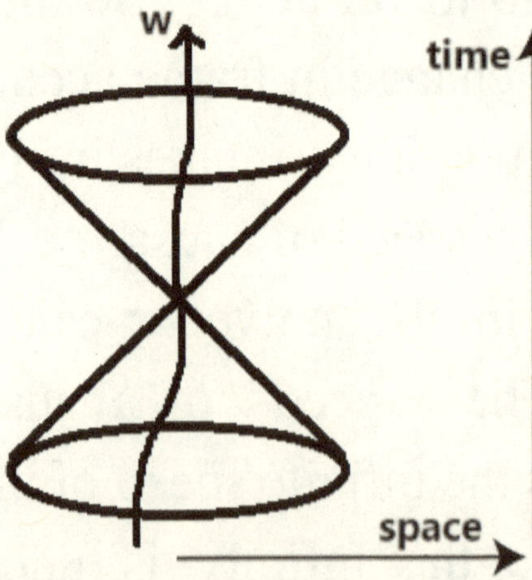

W is the worldline of a particle

When we talk about the Lorentz transformation, it's nothing but a change of frame that we use to study the motion in Minkowski spacetime. This transformation as discussed is analogous to the hyperbolic rotation. This transformation underlined by the concept of Minkowski spacetime, describes the motion of the universe. It predicts time dilation, length contraction, and many relativistic effects. The Lorentz transformation is something that is used in the special theory of relativity.

This chapter teaches one about the important characteristics and nature of spacetime. It shortly teaches how the spacetime changes for any reference frame that is nearing the speed of causality. In upcoming chapters, these concepts are taught in more detail.

Seventeen

The Universal Rest Frame of Reference

There are two frames of reference, inertial and non-inertial. In an inertial frame of reference, the moving train is not at rest. But in the frame of the passengers inside the train, they are at rest. And it is the world outside the train that is moving in the opposite direction to the train. In such a case, which frame of reference is the correct description of reality? This is one scenario.

In another scenario, let's have two twins. One of whom stays on earth, while the other takes a long trip within the universe. For the twin, on earth, it is his brother who is moving. But in the frame of reference of the twin inside the spaceship, it is his brother who is moving. In such a case, whom does time dilate for? Is there any universal rest frame that decides this? Let us understand this by still another scenario.

Let's imagine a series of planes where each plane sits on an adjacent plane. And where each plane shows an increment in speed. Starting with the first plane, the plane 1 is at rest and has a speed of 0 km/s. The plane above it (Plane 2) travels at 60,000 km/s which is 20 % of the speed of causality. But people living on this surface think that they all are at rest in their frame of reference. Plane 3 which lies on plane 2 travels at 60,000 km/s on plane 2. So, plane 3 is traveling at 1,20,000 km/s in the frame of reference of plane 1. Which is 40% percent of the speed of causality. But people on plane 3 think that they are at rest in their own frame of reference. Plane 4 which lies on plane 3, does the same and travel at

60,000 km/s on plane 3. Making it travel at 60% speed of causality in the frame of plane 1. Again people there think that they are at rest in their own frame of reference. Plane 5 does the same on plane 4. And plane 6 on plane 5. Now, even if people on plane 6 think that they are at rest in their own frame of reference, they are traveling at “c” in the frame of reference of plane 1. So, is it that easy to trick the universe? Is it really that easy to travel at “c”? The answer is solid no! Because there does exist a universal frame of reference. Where every velocity increase is counted, no matter what frame of reference you think you are at rest in.

In that way, the people on planes 4, 5, and plane 6 will start feeling the relativistic effects due to nearing “c”. Even if in their own frame of reference they are just traveling at 20% of “c”. The universal rest frame (Here Plane 1) counts the speed add-ups of every subsequent plane. So yes, there exists a universal rest frame.

Eighteen

Energy, Mass, and Relative Mass.

When we say energy, what does it exactly mean? Let's say the ball has a kinetic energy of 120 Joules. What does it mean? It means that the ball can carry out some work just by transferring its kinetic energy. Energy in physics is always associated with some form. When we say that a particular object has a particular energy, we mean how much energy it can transfer to any desired system to carry out some work. This is how we see energy. There are various forms of energy in this universe.

Just after the big bang, the only form the energy was in the form of Kinetic energy of massless particles traveling at "c". Today, there are many forms of energy like kinetic energy, Nuclear energy, chemical energy, electric energy, magnetic energy, etc. The total amount of energy in the universe is always constant. Now the question is "What does this "energy" look like?

The energy is like a JACKET that can only be seen when something wears it. And the only way to take this JACKET off is by transferring it from one system to the other.

Understanding the energy as the JACKET here as explained is important. The reader is requested to remember this for the content of this book ahead.

Just after the big bang, this JACKET was worn by the massless particles traveling at "c" in the form of a semi-permanent

JACKET called “frequency”. This was 10^-12 seconds before the big bang.

When the universe cooled still further, these jackets manifested as other semi-permanent and temporary jackets. After the universe cooled down, the Higgs field became active and particles started acquiring the mass. The Higgs field did nothing but convert one semi-permanent JACKET (frequency) to another semi-permanent JACKET (the rest mass). It’s called the rest mass. This mass is nothing but a form of the energy itself. Semi-permanent because it's transferable or convertible to other semi-permanent or temporary energy jackets. The rest mass can be converted to energy back again. Yes, I am talking about the famous mass-energy equivalence.

$$E = mc^2$$

In the next chapter, we are going to talk about this equivalence in detail. But before that lets have a short talk about the relative mass.

There is a difference between relativistic mass and the rest mass of a particle. The rest mass is a semi-permanent JACKET. But the relativistic mass is like a temporary JACKET. Kinetic energy, potential energy, electric energy, magnetic energy, etc are some of the examples of temporary jackets. Temporary because they all are based on semi-permanent jackets. Without the semi-permanent jackets, these quantities have no existence. So, the relativistic mass is a temporary JACKET. It doesn't exist if there is no semi-permanent JACKET of "Rest Mass".

Here, the titles Permanent, Semi-Permanent, and temporary names are mentioned to give an exclusive idea of the types of forms the energy takes in this universe. The above-mentioned temporary jackets are all based on Mass. Which itself is a semi-permanent JACKET. Another example of a semi-permanent energy JACKET is the frequency of Massless particles like a "photon". Though it has no mass, the energy is in the form of frequency. So the rest mass and frequency of massless particles are the semi-permanent jackets. In this universe, the only things we see and sense are the semi-permanent and temporary energy jackets. The Permanent Energy JACKET is something we discuss ahead in this book.

Ideally, when we talk about mass, we mean by quantity because we do know that the value of rest mass ideally doesn't change! By introducing one to the relativistic mass, one will forget that the mass ideally represents quantity. Let's take an example of an apple with weight of 300 grams. Ideally, this mass is looked at as the function of matter content in the apple. Now let's carry this apple at 90% speed of causality. And now the mass of the same apple becomes 688 grams with no increase in content inside the apple whatsoever! The extra mass added in this process is called the relativistic mass. When anything travels at "c", its mass starts increasing. The concept of relativistic mass makes one aware that the concept of mass is not a measure of the quantity of matter (Generally speaking). Now let's discuss why the mass increases at near to "c" speed.

Nineteen

The Energy-Mass Equivalence

Before we start let's remind the concept of energy as a JACKET for clarity:

The energy is like a jacket that can only be seen when something wears it. And the only way to take this jacket off is by transferring it from one system to the other.

The Rest mass is one of the forms of energy. We called rest mass the semipermanent JACKET. And in the last chapter, it was explained why we do so. The rest mass can be converted back to energy. This is what the famous equation says :

$$E = mc^2$$

Here E is the rest mass energy, m is the rest mass and c is the speed of causality.

So, since the rest mass can be converted to energy, does that mean that the Higgs field loses its meaning for the particle at and after that moment? The answer is no. The Higgs field still doesn’t lose its meaning. The Higgs field transfers the semi-permanent JACKET (frequency) to another semi-permanent JACKET (rest mass). So it is just a transfer of a JACKET. Because it's transferable or convertible to other semi-permanent or temporary energy jackets.

So the rest mass is basically a semi-permanent JACKET. Anything in this universe with rest mass can never travel at “c”. If he tries to do so, then the energy (which is a JACKET that can only be got off by transferring it from one system to another) adds up in the form of relativistic mass (which is a temporary JACKET like kinetic energy, potential energy, etc). And so, as the mass adds up, the more energy (JACKET) we require to add into the system. To achieve the speed of causality “c”, it requires infinite energy to transfer. In the case of any particle, the energy required to accelerate to “c” would be much more than the energy that can be harnessed from its rest mass. Since it takes infinite energy to accelerate a rest mass to the speed of causality, it is looked as impossible. Which is in fact the case.

Mass is energy (A semi-permanent JACKET), we know that. But the main question is “Why the speed of causality in the equation?”. Take another look at the same equation. The E is the rest mass energy, “m" is the rest mass, and “c” is the speed of causality. The stronger the particle interacts with the Higgs field, the more is its rest mass. And so, more is its rest mass energy. But what relates this rest mass energy to the speed of causality?

Just after the Big Bang, when Higgs field wasn't active, all the particles were massless. Travelling at "c" in space and at rest in time. But after the universe cooled down, the higgs field gave rest masses to these particles, making these particles unable to travel at "c". Matter can be devided into two types in this universe :

Case 1) The ones that travel between speed range of 0 to "c", but never at "c". This is the case of particles with rest mass.

$$\mathbf{c > v \geq 0}$$

c is the speed of causality
v is the speed of the particle

Case 2) The ones that travel only at "c". This is the case of particles that carry frequency (no rest mass).

c = u

c is the speed of causality
u is the speed of the "massless" particle

Just after the big bang, case 2 was the only case. But after the Higgs field became active the semi-permanent JACKET of frequency was transferred to the semi-permanent JACKET of rest mass, the particles got diverged into the two cases.

After interacting with the Higgs Field, many particles transmitted from case 2 to case 1. For the transition from case 1 to case 2 takes infinite energy. It is the jump from case 2 to case 1 that gave rise to the concept of rest mass energy. The more strongly the particle has interacted with the Higgs field, the more is its rest mass. So the rest mass-energy is actually the mass. But the question is "Why is there a speed of causality in the equation and no other number?".

Just after the Big Bang, all particles traveled at the speed of causality. Later, when some selected particles interacted with the Higgs field, the only speed they could travel was between 0 to close to "c". In reality, 0% of "c" and 99.9999% of "c" are equally close to "c". Since "c" represents the infinite speed of the universe, anyone travelling at 99.9999% of "c" is still as same as travelling at 0% of "c". Except that there are extreme spacetime dilations for speeds near "c". So, there is no way to accelerate a particle to "c". So, even if the equations of rest mass-energy and kinetic energy look alike, both of them are different forms of energy. They are just totally different JACKETs. When the Higgs field was inactive, the energy JACKET was in the form of frequency. After the interaction with the Higgs Field, the semi-permanent JACKET of frequency was transferred to the semi-permanent JACKET of rest mass. I am mentioning the JACKET analogy on a repeated basis as I want the reader to know the fundamental concept of energy. So, from traveling always at "c" to never at "c" is a solid change that manifests into the rest mass-energy of a particle. This is up to bottom way to know the role of "c" in E = mc^2. Up to bottom because we are going from

c = u

to

c > v

u and v are speeds before and after the particle interacted with the Higgs Field. Different notations are mentioned to note this difference.

This is the approach from frequency to the rest mass.

The Lorentz transformation makes sure that "c" is the infinite speed of this universe. The work-energy theorem is based only for the particles with rest mass. When we derive the total energy by applying the work-energy theorem with Lorentz transformation for any particle with non-zero rest mass, we discover that its total energy is not zero when it's at rest. Instead, it's huge number. It's an energy value so huge that it can accelerate the same given mass to 86.6% of c. For any given rest mass, the value of this energy is mc^2. This is the bottom to up way to know the role of "c" in E = mc^2. Bottom to up because we are working here in a range from

$$c > v \geq 0$$

(from 0 to c)

Knowing that the "c" in mc^2 needs both of these approaches (up to bottom and bottom to up). Up to bottom approach explains the impact of transition from travelling only at "c" to always below "c". While, the bottom to up approach explains how the approach of the universe to make "c" infinite and maximum makes developed equations unveil that rest mass carries an enormous amount of energy mc^2. A particle receiving mass, is not dependent on relativity. Receiving the rest mass is just the transfer of JACKET of frequency to JACKET of rest mass. But the rest mass-energy that accounts for c^2 in it is what the theory of relativity responsible for. And both of the above mentioned approaches collaborate to explain this effect of relativity.

The transfer of the JACKET of frequency to the JACKET of mass happens after the Higgs Interaction. One can imagine it as the Higgs field itself is taking bite out of energy where every bite is a multiple of c^2. The size of the bite is related to how strongly the particle interacts with the Higgs field, but it always is the multiple of c^2. And the spacetime dilation makes it sure that it always is c^2!

And why does spacetime dilate? Just to keep the speed of causality infinite! Thus the c^2 in the mc^2 is explained.

For those who seek very easy language:

The speed of light is the infinite speed of the universe. Here infinity actually means the end of the numbers (in terms of speed). So it is the real maximum. The universe makes this number infinite through spacetime dilations. Just imagine that the number line is in your hands and you define what is infinite. Things that travel at "c" are always in the form of frequency. The frequency is a semi-permanent JACKET. When these particles interacted with the Higgs field, the JACKET was transferred in the form of another semi-permanent JACKET called "The Rest Mass". This rest mass but always travel in between 0 to c speed (But never at "c") in space.

As it gets near to "c" in space, the spacetime dilation comes into the action to make "c" infinite. This spacetime dilation also makes sure that when the rest mass is *at rest in space, it carries an energy which is proportional to "c^2". The spacetime dilates for "c" and it makes sure that this is the number associated with the rest mass-energy of any rest mass. Thus it's just the dynamics of the same Lorentz transformations. The existence of rest mass energy isn't dependent on relativity. But the value of rest mass-energy is!*

Hence, the "c" in energy equivalence.

I won't include math in this book. But when one actually checks some maths of relativity, the idea will become more clear.

Twenty

The Massless Particles

The massless particles are the particles with no rest mass. For example a photon and a graviton. These particles can never travel below “c”

$$\mathbf{c = u}$$

c is the speed of causality
u is the speed of a massless particle

Provided, they have no rest mass, their rest mass energy is zero. This massless particles but still are one of the jackets. And the value of the JACKET can be calculated by it’s frequency.

These are the particles that travel at the infinite speed of the universe, aka the speed of causality “c”. Since, they travel at the fastest speed possible in the universe, in their frame of reference, space contracts to 0 length and time stops for these particles. This is even true in our frame of reference. Anything that travels near to “c” appears shrunk in length. In that sense, the light must shrink to 0 length in front of our eyes! But still, how do we see them existing? And if time doesn’t exist for light, why it still take 8 minutes to cover the distance between the earth and the sun if it’s omnipresent throughout spacetime? We will see that in the chapters ahead

Twenty-one

One's Own Frame Of Reference

In special relativity, where there is change in speed, there is relativity. Two frames or reference at same speed agree to each others measurements, Different frames of reference dont. No matter what speed one is travelling at, one is always at rest in his own frame of reference. Matter needs spacetime to exist. And what distinguishes space from time is the behavior of matter in it. So, even in the frame of reference that travels at "c", spacetime still does exists in it's own frame of reference. So even if light's spacetime shrinks down to zero in an inertial frame of reference, the spacetime does exist in light own frame of reference. But from the light's own perspective it is the entire universe's spacetime that appears shrunk.

So at the slowest to infinite speed of this universe "c", the spacetime remains intact. So light's personal spacetime stays intact. And we thus see it existing.

This concept of one's own frame of reference needed a special chapter as it's a concept one needs to keep in mind, while dealing with various spacetime curvatures throughout the theory of relativity.

Twenty-two

The Spacetime Of a Photon

The photon or light travels at the infinite speed of the universe “c”. So, in the inertial frame of reference, the universe outside has shrunk to zero length and time doesn’t run for the photon. In this sense, the photon must be omnipresent throughout the spacetime. But still, why does it take 8 minutes for a photon to reach Earth from the sun? Think!

To understand this let’s take a spacetime perspective. We might think that we can be rest as per our will. But that’s in space or in time, but not both. We always keep moving at the speed of causality in spacetime. The reader after reaching this point of the book knows it very well. When one rests in space one moves in time and when one rests in time, one moves in space. Moving at "c" in spacetime, we create a trajectory. This trajectory is called a worldline.

The speed of causality is the speed of existence. It is this speed that draws the worldlines in spacetime. So matter-energy of all sorts have a worldline associated with them. The direction of this worldline depends on how one distributes "c" in time and space.

In relativity, we track the worldlines in Minkowski spacetime, where light travels at an angle of 45 degrees. Since light travels at the speed of causality, nothing in the universe that travels below "c" can have its worldline tilted more than 45 degrees.

In reality, from an inertial frame of reference, the light never travels in this spacetime. Since spacetime is shrunk to zero length for a photon, it's omnipresent along the 45 degrees as soon as it is created.

Look at the diagram below. Let's take a scenario where 3 consecutive photons P1, P2, and P3 have been released from the sun in a spacetime diagram. Since the spacetime has been contracted to zero for light, its worldline marks a solid line in the spacetime. Which is omnipresent there. The "W" here represents the worldline of an inertial frame. We are in this frame of reference. The worldline as seen in the diagram doesn't trace the solid line as one can see, the future evolution is shown dotted here. We trace this worldline at the speed of causality "c".

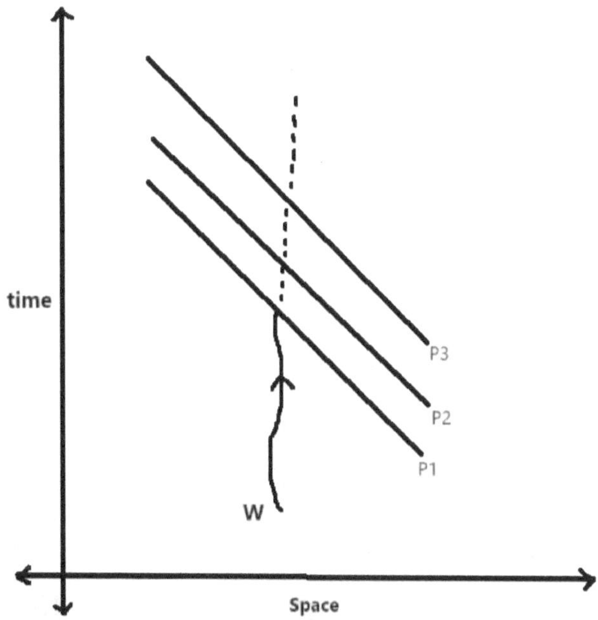

As one can see in the image, that light is omnipresent throughout the spacetime. And that's we who are intersecting the worldlines of omnipresent photons P1, P2 and P3. And since we do this at "c", we see each photon as if it's physically travelling in space to reach Earth in 8 minutes after it is released from the sun at the speed of "c". So, in our frame we see light travelling at "c" not because it is doing it. The omnipresent light appears to do so because it is us who are meeting it at “c” in time. So, we meet each of the omnipresent photons at “c” in time. This appears as if the photon is travelling in space not being omnipresent, which is not quite true.

So answering the main question, the light emitted from the sun is omnipresent throughout spacetime. And it's we who are meeting them at “c” in time. Let's take one more example. Two friends stand at a distance of 1 AU (Astronomical Unit) from each other. One of the friends shines the laser towards another friend. The laser becomes omnipresent in spacetime. It takes 8 minutes for the laser to be seen by the other friend not because it takes the laser 8 minutes to travel the distance. But instead, it’s the friend who takes 8 minutes to reach the omnipresent light of the laser. As the friend moves at “c” in time.

We all move in time at "c". This is the speed of causality (cause and effect). It is the speed the world evolves. So remember, when lights appear to travel at "c", it's you who is moving through the light at "c" in time. The light can be considered as the tress on a highway, and you as a car moving at "c". The trees appear to move in opposite directions at the same speed you are moving but in opposite directions. Similarly, the light appears to move at "c", but instead it is you who is moving through the light at "c" in time.

Twenty-three

The Notion Of Simultaneity

In various frames of reference, the same event can occur at different points in spacetime for two observers. When two events separate in space but non-separate in time, they are called simultaneous events. It's explained that the universal rest frame of reference does exist. No matter what frame one is in, if two events are simultaneous in the universal rest frame, then they are called simultaneous events.

The line joining two simultaneous events in spacetime is called the line of simultaneity. Simultaneity can be relative. But the real simultaneous events are the ones that are confirmed to be simultaneous by the universal rest frame. When two events are not simultaneous they are called non-simultaneous events.

The Betelgeuse Star is 650 lightyears away from us. So we take 650 years to reach any photon of Betelgeuse (Yes, I mean that!), once it's produced. And so we see Betelgeuse how it was 650 years ago. But what's happening on the Betelgeuse "right now" in the present is what the line of simultaneity connects. The concept of simultaneity is not a new concept if the reader is ever a physics student. But it was important to revise it, for the application in the upcoming chapters.

Twenty-four

The Lightcone

A lightcone is an imaginary boundary around an observer or anything that traces a worldline in Minkowski Spacetime. Light travels at the speed of causality, which is an infinite speed of this universe. It always travels at 45 degrees in Minkowski Spacetime. When we add one more spatial dimension in Minkowski spacetime, we get a cone out of this trajectory traced by a light. This cone marks how steep a word line of anything that travels below "c" in spacetime can go. Which never goes more than 45 degrees, since 45 degrees marks the speed of causality "c" (the infinite and maximum speed achievable in this universe).

The below image shows the lightcone around the worldline (w) of a rocket that moves slower than "c" :

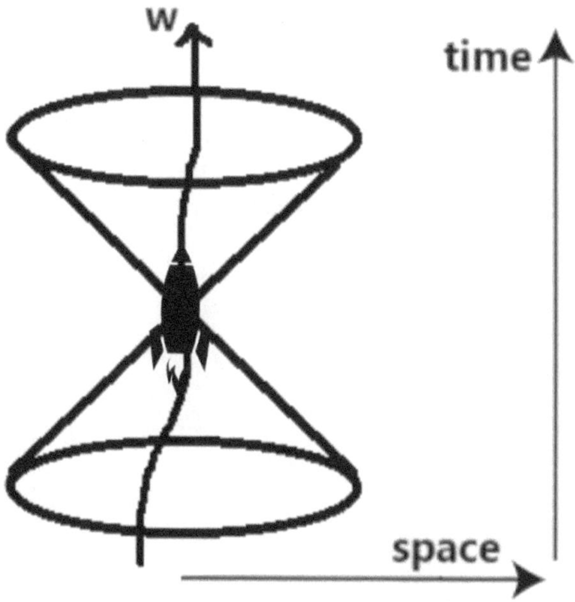

W is the world line of a rocket. The sides of a cone are tilted at 45 degrees angle.

For any successful communication, the information shared must lie inside one's lightcone. Or else, the information would be dodged! In the universe, most of the information shared is in the form of light. And so any information shared through light communicates through the edges of a lightcone. The worldline tilts when the traveler accelerates in spacetime. This brings the relativity in the information received. So the information can literally be dodged as if it never existed in the spacetime. The next chapter explains this.

Twenty-five

The Horizon of Special Relativity

The horizon is a boundary between two events that make them not communicable temporarily or permanently. In this chapter let's learn about the horizon of special relativity.

Two friends Alice and Bob initially at rest in space watching their favorite star Betelgeuse while tracing their respective worldlines within the spacetime. Later, Bob decided to accelerate at a constant acceleration for infinite time (provided there is an infinite energy supply) tracing the curved worldline of a rectangular Hyperbola. And so tilting his worldline from the initial rest frame of reference. To Bob's astonishment, after a particular speed gain, and in constant acceleration, a point comes where he is unable to see the Betelgeuse. He communicates with Alice (who is at rest), If everything is right with the Betelgeuse Star. The Alice, confirms the existence of Betelgeuse Star. But Betelgeuse is still not visible to Bob. Later when Bob slowed down in speed, he started to see the Betelgeuse again! Bob verified this multiple times.

So what's really happening in this case? Nothing but, this is the effect of the horizon of special relativity between the Betelguese and Bob! Which blocks the light of Betelgeuse to reach Bob. In reality, the light beams are omnipresent, and it's Bob's choice to dodge them in spacetime. With no will to slow down, it creates a temporary horizon between Betelegeuse's light and

Bob. This horizon of special relativity is called the “Rindler Horizon”.
This is better understood on the diagram:

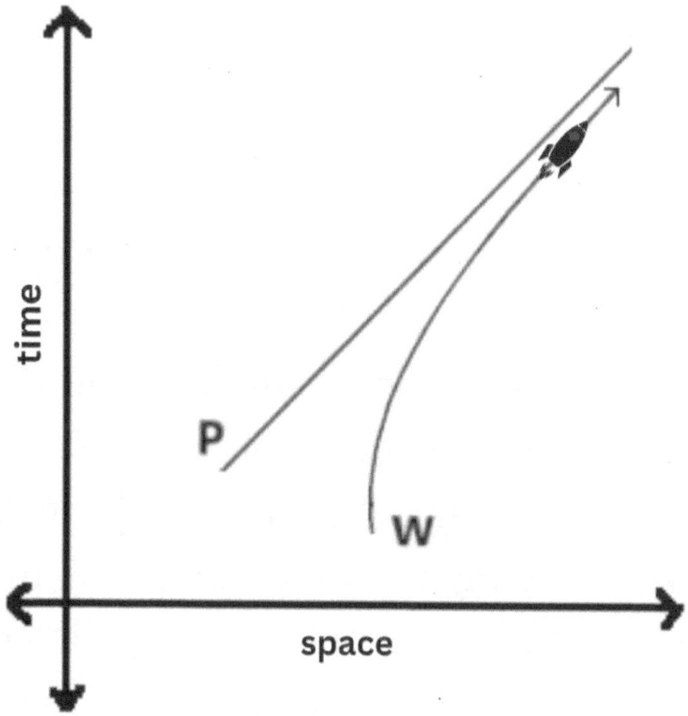

Here W is the world line of rocket (with Bob inside) under constant acceleration for infinite time. And P is the omnipresent worldine of the photon from the Betelgeuse.

As one can see in the image, as soon as the rocket is in constant acceleration for an infinite time, it dodges (barely here) the P. So the photon from the Betelgeuse remains uncommunicated as soon as the rocket is in constant acceleration. With any slow down or stop, Bob inside the rocket can see the Betelgeuse again. The constant acceleration leads W to dodge the omnipresent P. This makes it appear that there is some boundary that stops the light from reaching Bob from the Betelgeuse. This boundary is called the Rindler Horizon. The Horizon of Special Relativity.

Twenty-six

How We Think Of Time

Time is a dimension like space. What makes it different is the behavior of matter with respect to space. When one speeds up in space, one slows down in time. And when one slows down in space, one speeds up in time.

There are various notions of time. The most commonly known notion of time is that it preserves the memories or the information. As humans, we continuously grab the information. When we think of the past we call it the "experience", when we feel the present we call it a reality and when think of the future we call it an idea. The commonly believed notion is that the "time" records all of this information. While one believes that past, present, and future are predetermined.

Another notion of time talks about the many-fingered nature of time. This is the idea of time in the quantum many worlds theorem. Which says that the future is many-fingered and that multiple realities do exist.

Another notion of time says that the past and future are just information and the only thing real is the present. It says that time as a dimension like space preserves no information. And that the concept of past and future are nothing different than a piece of information. And that the only thing real is the present. This notion of time is called Presentism. This notion of time is what I believe. And it deserves a special chapter.

Twenty-seven

The Presentism

Presentism is a concept that says that the only event possible in the universe is the event of the present. But how does it match with how we actually perceive the time? When we think of time we refer to the creator and creation. We think that the past creates results in the present and the present results to the future. But now I want a reader to take a look on a time dimension which is the same as a dimension of "space".

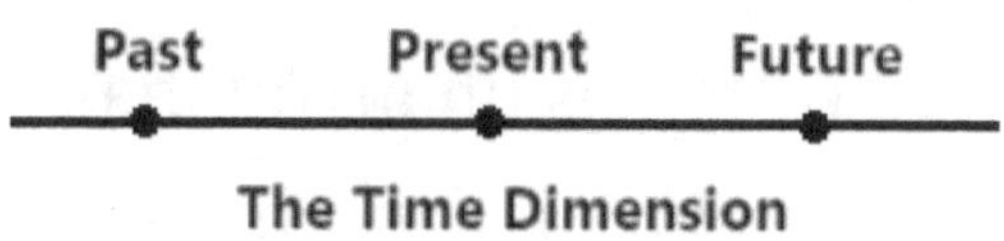

The past, present, and future are the points located in the time dimension. The Past is the state of the universe, the present is the state of the universe and the future is the state of the universe. Thinking this way why do we think that the past is the creator of the present? This question is not silly! Why can't the present be the creator of the past referring to the points on the time dimension? As directions (left or right) are of equal importance in space, the same applies to time. Then why have we labeled past as the creator and the present as the creation?

The entropy? Yes! The second law of thermodynamics says that the entropy of the universe always increases. We have assumed that time has an arrow based on the increasing entropy of the universe. But in reality, the time has no specified direction or the arrow. In fact, no dimension has! Just as we have given time the special arrow based on the second law of thermodynamics, any quantity can be given the arrow. Even for scalar quantities like temperature. Plotting anything on a graph doesn't give any special arrow to that quantity. In many diagrams mentioned before in this book, the time dimension has been given an arrow. But that doesn't mean that it really has! The direction of time there just represents the direction of our motion in time. Time is just a dimensions as "space".

Another thing, do you think if the entropy is going to decrease if one moves in the opposite direction in time? The answer is solid no! Reversing the direction of how we move in time is not the same as reversing the video on a mobile phone. In fact, the entropy would still be increasing.

In the universe, the entropy always increases from low entropy to high entropy.

And labelling the low entropy as the creator and high entropy as the creation doesn't make sense. So, labelling the past as the creator and the present as the creation in the dimension of time based on entropy is wrong. In fact, when the present creates the past, it's called the "creation under the cause of the present" (Well, discussed in my book, the next level observer effect). Time must be thought of as a dimension in space. Where things at rest in space travel at "c" in time. The increasing entropy increases at "c" for something that rests in space. Entropy is linked to matter-energy's evolution in time. But time itself has no obligations with entropy. The entropy is the function of time. But time is not the function of entropy.

There are three things in the universe, the present we feel, the experience of the past, and the idea of the future. The experience and idea come under the domain of information that we carry while feeling the present. Otherwise, the universe doesn't count the past and future, what counts is just a current state which is nothing but the present, and the rest of everything is just the information (There exists the universal rest frame, the present we are talking about is based on that. So there are no contradictions from relativity (to mention)).

The sunlight takes 8 minutes to reach the Earth. Now, doesn't that mean that we are looking at the sun's past? To believe the present as the only real thing and the past as only information may be tough initially, but after constant hammering, one can actually understand it and start believing in it. Now, even if that shows the sun 8 minutes ago, it's just a piece of information that we read in the present and nothing more. So, the only thing in the universe is the event of the present! This is the conclusion of this chapter.

Twenty-eight

Time Travel

When one speeds up in space, one slows down in time, and when one slows down in space, one speeds up in time. This effect is only considerable at speeds comparable to the speed of light. So in that sense we all are travelling almost at full speed in the time dimension. I mean at “c”. Can we call this time travel? I mean, why not? This is in fact a time travel.

Well, if one slows down in time, the apparent effect can be as if one has travelled in time. For example, the twin paradox. Where one of the twins stays on Earth and the other takes a long trip within the universe at speeds near to “c”. And when the twins meet again on Earth in one reference frame, the twin on Earth would be older than the twin who was on a trip. The younger twin slowed down in time while speeding up in space. And yes this can be considered as time travel. A travel that comes by slowing the speed down, the time travel! And apparently we find ourselves in the future. But unlike the sci-fi movies, one will not find their future self here. The science fiction movies assume a predeterministic time. In other words, it's based on the belief that the past, present and future are already determined. By reading the previous chapter, it's easy to conclude that such assumptions are baseless. Also we have a practical proof of it. If one slows down in time and goes into the future, one doesn't find the future self. We keep accelerating the particles like electrons at close to "c" in the particle accelerators and still have never seen the present particle and the time-traveled electron in one frame. This is the proof!

Also, the time travel in the past still wouldn't be as same as the sci-fi movies. If one time travels in past and meets their old self, one adds an extra person to the population, which is not how actual time travel in the past would look like. Also, to reverse the direction of time, it would take faster than "c" speeds. Which is indeed not possible in this universe.

Now, even if one reverses the direction one is moving in time, it will make no difference! Everything would be as same as it normally was. The universe doesn't preserve the memories (the concept of presentism). This is about the time travel.

Twenty-nine

Spacetime and Motion

When we move in spacetime, the spacetime changes coordinates for us changes and we call it motion, but when we go through the disturbed spacetime, will we move? The answer is yes! We have control on our motion in spacetime (there is a tiny exception to this). If one wants to slow down in time, one can do it by speeding up in space. And if one wants to speed up in time, one needs to slow down in space. Throughout the theory of special relativity, we have discussed all of these. And it was all our choice.

When an object at rest goes through a curved spacetime, it starts to move in space.

Why ? The answer is simple! An object at rest moves at “c” in time. And when it goes through a curved spacetime, it’s motion in time is further continued. But now the curvature in time has slowed down its speed in time as comprared to the flat spacetime. And when one slows down in time, one speeds up in space. So when an object at rest or anything travelling below “c” goes through the spacetime curvature, it gains some speed in space.

And what creates this spacetime distortion or spacetime curvature is mass!

Mass tells spacetime how to curve and spacetime tells mass how to move.

The speed gained in space by this spacetime curvature keeps on increasing as one nears the source. This acceleration is nothing but gravity!

The mass of the Earth tells spacetime how to curve, and the same spacetime curvature tells the satellites of Earth and everything how to move. The more massive something is, the more the spacetime curvature is created by it. Even normal day objects curve the spacetime, but the magnitude is too tiny to consider.

First, we understood the nature of spacetime and how it differs from hyperspace. Then we talked about how the spacetime transforms for anyone or anything traveling closer to "c" (Special Relativity). And now we are at the theory of relativity, which is more general.

The special relativity comes from motion that is subject to one's choice. A set of things can be at rest and others in motion, which is choice-based. So it's one's special case to deal with. Thus, the special theory of relativity. While the general theory of relativity talks about relativity on a wider scale, making it general. As we get ahead in this book, it will make more and more sense why this theory is the general one.

Thirty

Spacetime Curvature With Respect to What?

Mass tells spacetime how to curve or distort, and spacetime tells mass how to move. The concept of spacetime comes from the asymmetrical behavior of matter in spacetime. So, there is no concept of spacetime if matter doesn't exist. So every spacetime point in the universe is accompanied by matter-energy. Also, the mere existence is supported by spacetime. So, spacetime and matter do co-exist.

When we say spacetime curvature, we have to first define flat spacetime as the benchmark. Flat spacetime doesn't mean zero matter-energy. Throughout the universe, there is something called the 0-point energy. This energy defines the flat spacetime and the spacetime itself! The 0 energy is not possible anywhere in this universe!

Throughout this book, whenever we call for the spacetime curvature, it would be with respect to this flat spacetime, which has non-zero energy density (the zero-point energy).

The Energy Conservation.

The energy can neither be created nor destroyed. That's the very basic principle of physics. But no one might had mentioned it that it was for the limited scale of the universe. Globally, the energy is not conserved!

Thirty-one

The Mass-Energy Density.

The general theory of relativity teaches us how an object with mass creates the spacetime curvature and how the same mass follows it. In case of two indivisible particles, one as massive as Earth and the other as massive as a grain of sand, both create their own spacetime curvature. The particle with the mass of a grain of sand will follow the spacetime curvature of the particle with Earth's mass, and the particle with Earth's mass will follow the spacetime curvature created by the particle of a grain of sand. But the spacetime curvature created by the grain of sand mass is much less even to consider other than zero. So it's the grain of a sand mass particle that follows the spacetime curvature of an earth mass particle. This is very basic.

In reality, there is no indivisible particle as massive as the Earth. But we can treat Earth like an indivisible system to consider its spacetime curvature. This system is possible because of the same force of gravity that assembled the Earth's material. And the electromagnetic forces that hold the material of Earth together, the nuclear force keeps the atoms stable. In the universe, there are many systems like these. Some of these systems are held together primarily by electromagnetic force, some are primarily maintained by the nuclear force, and some by quantum mechanical degeneracy pressure. This topic is important to consider as it talks about another important quantity, "The Mass-energy density". Mass curving the spacetime is not enough to describe the spacetime curvature for objects like stars, neutron stars, white dwarfs, etc. Generally speaking, mass-energy density curves the spacetime and tells it how to curve. For example, there is a huge difference between the spacetime curvature of Earth and earth Earth-sized neutron star. This is explained by the difference in mass-energy density of the neutron star and the Earth. The mass-energy density ranges from 0 to infinity in magnitude. And yes, infinite mass-energy density means infinitely strong spacetime curvature. This gear in spacetime can lead us to bizarre conclusions (later discussed in this book)

When spacetime distorts, time dilates. In special relativity, to bring out the time dilation (not curvature) for a normal day object, it takes a huge amount of energy. As one needs to travel closer to "c" to do this. On the other hand, the spacetime dilation in general relativity is quite plentiful and accessible in the universe. For example, if one wants their time dilated, then just go near a star close enough and that's it!

Thirty-two

The Fundamental Motion

From the absolute rest frame of reference, all the matter in the universe has been carrying a fundamental motion with it from its very birth. It marks its relation with spacetime.

The absolute rest in spacetime as a whole is not possible.

One has to move either in time or in space or fractionally both at the speed of causality. This is the fundamental motion of matter. This motion is what separates spacetime from the hyperspace. Other than this, there is no motion in this universe. When a particle is at rest in space, it moves at "c" in time. And when it moves at “c” in space, it rests in time. Besides this, there is no motion. When a particle is accelerated to a particular speed, its speed in the time dimension decreases. This happens until the particle reaches “c” in space. So, this is just an exchange of speed from space to time and time to space with max speed “c”. Anything travelling at any given speed in space comes to a deduction of its speed in the time dimension.

When the object moves through the curved spacetime, the same exchange happens. The curvature in spacetime automatically makes a particle’s speed exchange from time to space. And then we see the object moving in space. We call it gravity. Gravity is due to this speed exchange.

No matter how heavy the indivisible particle is, every particle under the same frame of gravitational influence accelerate at the same rate. Let's see why this is so.

Thirty-three

Why the Same Acceleration?

Two particles with unequal mass in the same spacetime curvature experience the same magnitude of spacetime curvature. It's easy to predict and notice this in the equations that the acceleration due to gravity is not dependent on the object's mass. But we need to know how it practically works.

We all move at "c" in time when we are at rest in space. Two indivisible particles with unequal masses starting off from the same height above Earth's surface fall at the same time. They accelerate at the same rate regardless of their mass. This is because they share the same spacetime curvature. The speed exchange from time to space for both the particles is the same. So both of them travelling at "c" in time exchange equal portions of speeds in space at any given time. Thus, accelerating at a constant rate.

When one drops an object down from a certain height, the speed exchange occurs from time to space. Considering the earth as purely spherical, as the object moves nearer to the surface of the Earth, the more the spacetime curvature felt by it. But as it goes below the surface and nears the centre, the spacetime curvature becomes less strong until it becomes zero at the centre. As the object goes below the Earth's surface, the spacetime curvature of the mass of the Earth left behind starts to nullify the spacetime curvature ahead of it. This happens until the spacetime curvatures of both hemispheres of the Earth nullify each other to create zero spacetime curvature at the centre.

So once the object crosses the surface of the Earth, the exchange of speed occurs in the reverse direction, i.e. from space to time. Until it returns to its fullest speed in time at the centre.

Spacetime Curvature also marks its omnipresence like light throughout the spacetime. And it's we who encounter it at the speed of causality in spacetime. But how? Let's see that in the next chapter.

Thirty-four

Encountering and Escaping the Gravity

Whenever something globally changes the structure of spacetime in whatever way, the subject of gravitation comes in. But how does it operate? Gravity, like light, travels at “c”. So, in the frame of a graviton, the entire spacetime continuum shrinks to a point. Making it omnipresent like light. Recall how we interact with light (page no.117) Look at the image below:

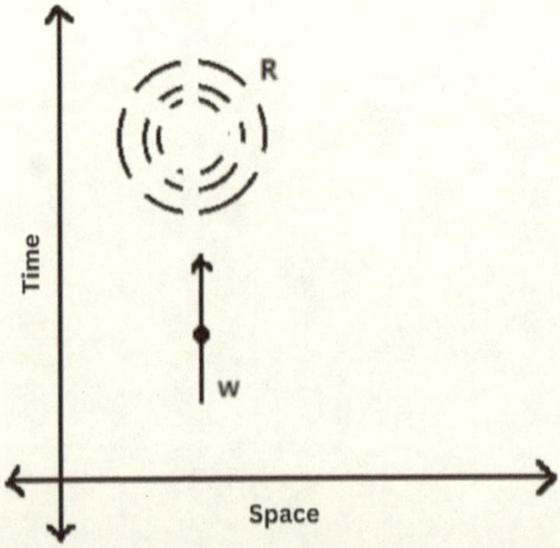

Here, W is the worldline of the particle and R is the spacetime curvature at any given event.

So, it's we who are travelling at "c" in spacetime and encounter the omnipresent spacetime curvature. As soon as one stays within this source of spacetime curvature, one keeps encountering it at "c". The gravity in our frame must be omnipresent. The only reason we see it travelling at "c" is because we encounter it at "c." This is how we interact gravitationally with the Earth and every source of gravity.

When the object at rest encounters the spacetime curvature, its trajectory in spacetime changes. To return to its original trajectory, it needs to be accelerated. The velocity needed to be attained by acceleration must be more than escape velocity for any particular source of gravity.

For anything on the surface of the Earth, the escape velocity is approximately 11.2 km/s. Anything that attains a speed greater than this speed will pierce the spacetime curvature as if the spacetime curvature didn't exist. Everything on the Earth's surface has its worldline curved due to the spacetime curvature of the Earth. So everything, no matter how massive it is, has the same value of escape velocity here on the surface of the Earth. To attain the escape velocity is our will to exchange the speed from space to time.

For example, anything here on Earth that travels in any direction at 11.2 km/s crosses the Earth's gravitational influence like it never existed. Things can go through Earth's atmosphere near to ground and back again to space if they travel faster than this velocity, as if the space-time curvature were never there.

Thirty-five

From 0 to c by Gravity

Let's take an example of a gravitational source which has an infinitely huge gravitational range with acceleration of gravity 10 km/s^2 (This is just an example) at an infinite distance "d" from the source. This source in this example is purely imaginary. An object starts accelerating to higher and higher velocities gravitationally. Until it reaches 99.999% of the speed of causality "c". And the object is still subjected to the acceleration. In such a case, will it reach "c"? The answer is no. As the object moves closer to "c", its mass starts increasing. This mass is called the relativistic mass (discussed earlier). And it keeps on increasing infinitely as the object is accelerated closer to "c". Theoretically, it takes infinite energy to do this. But the question that arises is, where does this energy come from? When we burn fossil fuel, we know that the energy comes from the electrochemical energy in the fossil fuel. But when an object is accelerated due to gravity, here is case to near "c", where does so much energy come from? The answer is the mass of the source itself! But in this case, this doesn't cause any mass loss to the source. Unless and until the object doesn't steals this energy and escapes the system. So even with the forever acceleration, the object can't reach the infinite speed of the universe, aka the speed of causality.

For anything that travels throughout any given spacetime curvature, its velocity at first increases and later, as it nears the centre, it decreases to zero. If the object is travelling faster than the escape velocity of the system, it will surpass the centre and it's speed again starts increasing until it reaches back to the surface and then again starts decreasing. At the end of the trip, it gives back the energy required to accelerate it at the first place.

Thirty-six

The Escape Velocity

The escape velocity of any system is the velocity travelling faster than which one literally ignores the gravitational influence of the system. When one travels faster than this velocity, one gets into the system, speeds up, slows down, speeds up again and then returns to the original speed, escaping the gravitational well.

The idea is simple: in general relativity, move in the curved spacetime. By counterbalancing this spacetime curvature locally, one can pierce their way through the gravitational well. The counterbalancing can be done by changing one's spacetime coordinates locally. This change in spacetime coordinates needed to counterbalance is nothing but the escape velocity. To escape the system one needs to travel faster than this speed.

On Earth, for anything to counterbalance Earth's spacetime curvature, one needs to travel at 11.2 km/s. Just like -1 plus +1 is 0. The speed 11.2 km/s cancels the spacetime curvature of things on the surface of the Earth. Equal in magnitude, but counterbalancing. Here on Earth, sitting at rest in the spacetime curvature of Earth is the same as travelling at 11.2 km/s in space! Same effect of time dilation. TRY CALCULATING IT FOR YOURSELF.

The dynamics of spacetime is to deal with the motion. When we rest in space, we travel at "c" in time, and when we travel at "c" in space, we rest in time. The change in spacetime coordinates locally in special relativity or going into curved spacetime of general relativity it is only the science of motion. The motion that never crosses the speed limit “c” in spacetime.

Standing here on Earth it is equivalent to travelling in space at 11.2 km/s. Being on the surface of any system is equivalent of travelling in space at the system’s respective escape velocity. This has a major advantage. Because there are systems in the universe that have escape velocity of “c”! Let's know about them.

Thirty-seven

The Escape Velocity "c"

There is an abundance of mass-energy systems in this universe, A star has a significant spacetime curvature. But the limits of how spacetime can be distorted don't stop there. The material of the giant stars tries to increase the mass-energy density of the star system by gravity. The only thing that keeps such stars stable is the strong nuclear force by the hydrophilic equilibrium. Once the star's primary nuclear fuel is over, the mass-energy density increases, and then the star starts to burn the secondary fuel and further. After all the fuel is fused into the heavier unfusable atoms, the mass-energy density increases until it either forms a white dwarf, a neutron star, or something that has the escape velocity “c”.

The speed of causality is the infinite speed of the universe. A gigantic star at the end of its life, with its huge mass, can create a mass-energy density that gives rise to the spacetime curvature so intense that the escape velocity itself becomes the infinite speed of the universe, the speed of causality. For anything with non-zero rest mass, it’s impossible to leave such a system. On the Minkowski spacetime diagram, anything in this situation will have a worldline tilted by 45 degrees like light. And the time totally stops for it.

Objects with such a mass-energy density are called Black Holes. The boundary of the black hole where the escape velocity

becomes "c" is called the event horizon of a black hole.
The spacetime doesn't allow the worldline of any object to tilt more than 45 degrees locally. In special relativity, we see that it takes an infinite amount of energy to cross the speed-of-light barrier. But the case of a black hole is something different. At the event horizon of a black hole, the worldline tilts at an angle of 45 degrees (in the frame of reference far away from the black hole). And the most astonishing is that it goes even beyond that! No particle with mass can attain this state in special relativity. In general relativity, particles with a non-zero mass can attain these worldlines.

Being on the surface of any system is equivalent to traveling at its escape velocity in space (same magnitude). We have learnt this in the preceding chapter. In the same way, being on the event horizon is equivalent of travelling at “c” in space. Even if travelling at “c” is not possible for a particle with non-zero mass due to special relativity.

Being at the event horizon of a black hole gives us an idea about what can hypothetically happen when anything crosses the “c” limit. This manifests as the world inside the event horizon of a black hole. Which is the famous unknown part of a Black Hole. To understand this world better, we use something called the Penrose diagram. The idea is the same as the Minkowski Space. Light travels at “c” and worldlines of any non-zero mass can’t tilt more than 45 degrees, etc. Except that now we are encapsulating the infinite spacetime into the picture which can be used to discuss any event of the universe.

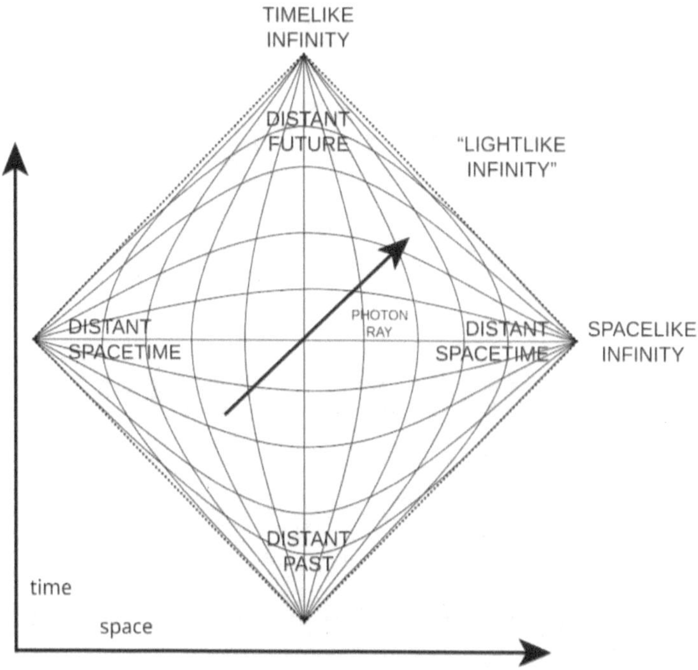

The image shows the very general case of the Penrose diagram.

The word Time Like differs from Spacelike by the ability of matter to behave asymmetrically In spacetime. As when one speeds up in space, one slows down in time and when one speeds up in time, one slows down In space

We will use this form of representation to understand the world beyond "c" in this book.

Thirty-eight

The World Beyond "c" and the Event Horizon

In many ways, the event horizon models the speed of causality. It gives us an idea about what can hypothetically happen when one crosses the "c" in special relativity. Before we move ahead, let's revise the nature of spacetime.

1) The Spacetime is a part of infinite-dimensional Hyperspace.

2) The time in spacetime is the same as the dimension of space. And it's a matter that makes time different by its asymmetric

behavior in one of the 4 dimensions of infinite-dimensional Hyperspace.

3) When one speeds up in space, one slows down in time and when one speeds up in time, one slows down in space

Above the event horizon of the black hole, the particle at rest travels at "c" in time and rests in space. And inside the event horizon, it's the same thing! Now let me elaborate if I have gone too fast. Again, it's not a typing mistake.

It is very clear that when one moves in space, one slows down in time, and when one slows down in space, one speeds up in time. At the event horizon of the black hole, time slows down to zero. Subsequently the speed must add up in space. For any massive particle, to be at the event horizon of a black hole and to travel at the speed of causality is not different at all. Even the worldline becomes tilted at 45 degrees like light. For such a particle, it's not possible to return to the universe outside. Since it takes faster than "c" travel for the particle to escape. Which is not possible due to special relativity. The frame of event horizon gives the massive particle a worldline tilt of 45 degrees like a photon, as if it's traveling at "c".

To travel at "c" either in space or in time or fractionally in both defines the underlying existence of matter in the fabric of spacetime and defines the concept of time in spacetime. But more than this, speed is never possible whatsoever according to special relativity.

When looked at from 3-space (that we can imagine), the event horizon is a 3-dimensional envelope of a black hole where particles with mass have their worldlines being tilted at 45 degrees as if they are travelling at "c". And they actually are!

The spacetime curvature doesn't stop at this but. Inside the event horizon, the spacetime curvature is more intense. When the particle is on the event horizon of a black hole, it moves at "c" in space and doesn't move in time. And just after crossing it, it begins from the absolute rest in space again! Yes! After one crosses the event horizon, the particle again starts with zero speed in space and at "c" in time. As if everything starts new.

After that, the spacetime curvature becomes so intense that matter simply doesn't exist! And we get back to hyperspace! This point of a black hole is called the "singularity". The new spacetime created between the event horizon and the singularity acts like a pseudo-universe. Or a temporary universe that breaks off at the singularity. Look at the Penrose diagrams below:

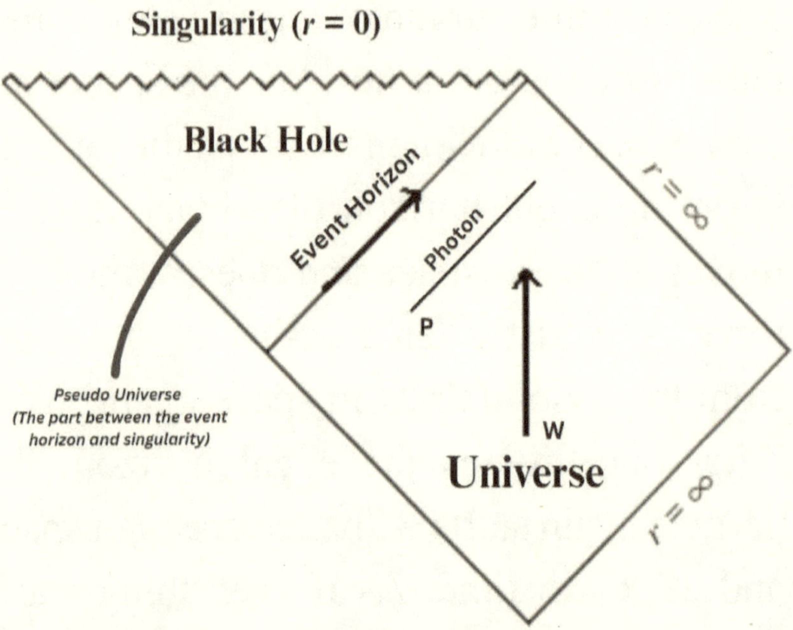

This is the representation of a Black hole on the Penrose Diagram,

The Event Horizon shows its omnipresence in the form of a boundary aligned at 45 degrees.

The wiggly line shows the break in spacetime. Which means the start of hyperspace.

The image below shows the difference between the worldlines of the particle at the rest frame, the particle on the event horizon, and the particle inside the event horizon (W, E, and W'):

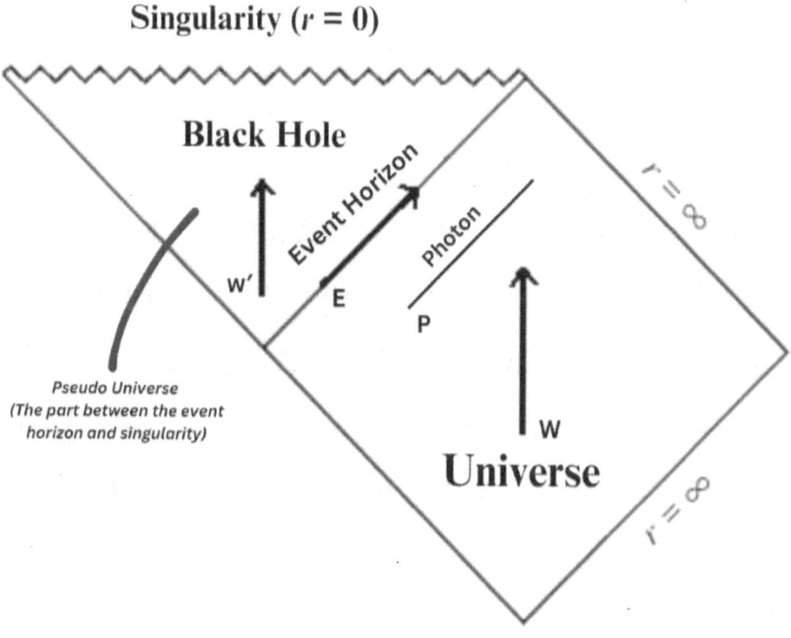

Here, W is the worldline of a particle at rest in the universe.

E is the worldline of a particle on the event horizon of a black hole. One can see that it is aligned at 45 degrees.

W′ is the worldline of the particle inside the Event Horizon of a Black Hole.

The pseudo-universe in the picture operates just like a normal universe. The space and time of spacetime appear to switch their roles when one crosses the event horizon of a black hole. The new spacetime operates just like the normal universe. Except that it is temporary, since it breaks off at the singularity. Once anything crosses the event horizon of a black hole, nothing can escape the singularity. As now in the pseudo-universe, the singularity has become a moment in time rather than a point in space. Thus can't be avoided. And the object has to inevitably meet the singularity.

As one crosses the event horizon of a black hole, a new spacetime (pseudo-universe) begins. The event horizon acts as a boundary between the universe and pseudo-universe. As one crosses the event horizon, the particle moves into the pseudo-universe to rest in space and travel at "c" in time and follow special relativity there.

The matter switches how it behaves in spacetime. Due to the switch in behavior of matter from resting in space and travelling at "c" in the universe (W) and travelling at "c" and rest in time at the event horizon (E) to back again to rest in space and travel at "c" in time in pseudo universe (W'), it appears that the spacetime has switched it's roles. And that spacetime itself flows inside a black hole. But it's just the game of behavior of matter instead.

Thirty-nine

Things On the Event Horizon

For a particle at the event horizon of a black hole, the particle rests in time and moves at "c" in space. This represents the worldline of a photon or anything that travels at "c". From the rest frame of reference, for anything that travels at "c" in space, the universe must shrink to an infinitely small dimension. Such a thing shows omnipresence throughout spacetime. For example, the light or a photon. Discussed earlier how we interact with these omnipresent things. Once when an object lays down over an event horizon, it's worldline becomes like a worldline of a photon. And it shows the omnipresence

(Discussed with examples in the next chapter).

The event horizon is a boundary between the pseudo-universe of a black hole and the universe. The things on the event horizon exist in both these realms. This gives us an idea about the existence of light. If there is a hypothetical world beyond "c", the existence of light would be somewhere in the middle of the universe, and the hypothetical world (pseudo-universe in case of a black hole).

Forty

Being Omni-present

There are two things in the universe: you and your experience of the events. A photon shows its omnipresence in spacetime. In the experience of a photon, it interacted with the observer as soon as it was created. This is the experience of a photon.

Being at the event horizon, it is the same as travelling at “c” in space. Travelling at “c” things become omnipresent (same as photons). Imagine a person is on the event horizon of a black hole at the centre of the galaxy. In this frame of reference, the milky way and Andromeda galaxy collision has already happened. And it happened as soon as he touched the event horizon. This is his experience. Not only the Milky Way and Andromeda collision, but the person with this frame of reference also sees the entire evolution of the galaxy and the universe in one scene. The person experiences all of these as soon as he touches the event horizon. For the surrounding universe, the person was always there from when he touched the event horizon till eternity.

The motion in time is all about the speed we experience the surrounding universe. At the fastest speed "c" in time, we know how things take place. But when traveling at zero speed in time, for him, there is no time gap between the cause and effect in the surrounding universe. Every experience is experienced all at once, no matter what time that experience dates back to. Resting in time is like being in hyperspace. There is no gap between the cause and effect.

Even if time stops for anything that is on the event horizon, the time for that object or traveler in its own frame just ticks as normal. This also applies to the photon (made it clear in the preceding chapters).

Forty-one

Beyond The Event Horizon

The event horizon of a black hole marks the boundary between the universe and the pseudo-universe. The word pseudo-universe is used because it marks the beginning of a new spacetime. New spacetime because the event horizon of a black hole represents a frame of reference of as if the particle is travelling at "c" and beyond that lies the pseudo universe, where the particle again starts with zero speed in 3-dimensional space and travels at "c" in time. It's not that the spacetime has switched the roles, it's instead the matter that has switched the behavior.

In case one finds himself in this pseudo-universe, one will find no difference between the universe and the pseudo-universe.

The difference becomes clear when one directly compares the pseudo-universe with the universe. The event horizon of a black hole represents the speed of the causality frame. The pseudo-universe that lies beyond the event horizon represents the (hypothetical) faster than “c” frame. This frame doesn’t come from accelerating beyond “c”. It instead comes from the behavior of matter-energy. When infinite energy is focused to accelerate something beyond “c”, the system forms a black hole. Until one realizes that the faster than “c” is nothing but back to zero. And less than zero is “c”. So, more than “c” is zero and less than 0 is “c”. To switch a particle at rest (zero speed) in the pseudo universe to "c" in the universe, and a particle at "c" in the universe to zero speed in the pseudo universe, it takes an infinite amount of energy.

Due to a switch in the behavior of matter-energy in spacetime, the "c" in space switched to "c" in time, so back to rest in space. The meaning we can take from this is that slower than absolute rest is "c" and faster than "c" is the absolute rest.

The pseudo-universe is the result of the switch in particle behavior in spacetime. On the event horizon, the particle was travelling at "c" in space and was at rest in time. But when one crosses the event horizon, the matter redefines spacetime into its zero speed in space and "c" in time, as if it comes back to rest in space again.

To make a direct comparison between the pseudo-universe and the universe, imagine a ball from pseudo-universe appears in front of Alice in this universe. The existence of the ball is supported by the spacetime of the pseudo-universe with no event horizon in between. Of course, it's a hypothetical scenario. Now you try to push this ball with your hand. And you find that no matter how large the force Alice applies, the ball will never move.

Suddenly, a boy from the pseudo-universe appears in front of Alice. The existence of a boy from the pseudo-universe is based on the spacetime of the pseudo-universe. The boy comes to take his ball back. The boy could easily lift the ball. Before the boy goes back, Alice describes his experience with the boy. And even the boy got stunned seeing this. Now, Alice tells the boy to lift the ball from her universe. Surprisingly, the boy couldn't lift it even after applying the huge force. The same experience from both the ends! I mean, what's happening here?

The pseudo-universe is the result of the switch in matter's behavior in spacetime. On the event horizon, matter was travelling at "c" in space and at rest in time (Special relativity equivalent). But when one crosses the event horizon, the matter redefines spacetime into its zero speed in space and "c" in time. This appears as a spacetime switch. When Alice tries to move the ball, she does so in space in her frame of reference. But in the ball's frame of reference, she is actually trying to move the ball more than the speed of causality in its time dimension. An effect of switched roles of space and time in spacetime. The same thing happens when a boy from the pseudo-universe tries to move Alice's ball in our universe. In the universe and as well as in the pseudo-universe, the speed of causality is the highest speed achievable, and nothing can overtake this speed either in space or in time! So, no matter that a very huge force is applied to move the ball in the pseudo universe by Alice and the boy from the pseudo universe, the ball will not move! In the pseudo universe, what is space for Alice is time for Bob, and what time for Bob is space for Alice.

The boy in the pseudo-universe could easily move the ball in the pseudo-universe because it's just like moving it in space with some non-relativistic speed in his frame of reference. This is the same as we play with the ball in our universe.

One more thing the two frames of reference notice is that they are not able to physically touch anything in each other's frames. The two frames of reference have totally switched coordinate systems. The only way things from one side can communicate to the other is through gravity and light. Alice and the boy from the pseudo-universe can see each other through light and interact through gravity.

This was just a hypothetical example to compare the spacetime and its switched version. In reality, the pseudo-universe always appears beneath the event horizon of a black hole. The main point of this example was to give the idea of the difference between the spacetime and its switched version. The closest we can get to this example is when Alice from our universe and Bob from the pseudo-universe meet in around the midway of a wormhole that connects the universe and pseudo-universe. In such a scenario, they can only interact through light and gravity.

Forty-two

Lightcones on and beyond the event horizon

Anything on the event horizon of a black hole and for anything traveling at the speed of light, the tilt in the worldline is at 45 degrees. When one crosses the event horizon of a black hole, one enters the pseudo-universe. Which has the spacetime with switched roles.

In the pseudo-universe, the same rules of special relativity apply (like our universe). With lightcone boundaries for the worldlines of things that exist in a pseudo-universe. Except one thing, topologically the lightcones in the pseudo-universe are tilted by 90 degrees as compared to the lightcones in our universe.

In any given universe (pseudo-universe or our universe), the light cones don't tilt. What tilts is the worldline. But once the event horizon is crossed, even the light cones tilt by 90 degrees, creating the special relativity boundaries for things that exist in the pseudo universe. On Penrose diagrams, this tilt isn't shown in the pseudo-universe, since as soon as the pseudo-universe starts we change our coordinate basis there. This eliminates the need to show the 90-degree tilt in the lightcone. This is how flexible the Penrose diagrams are. But in the end, it leads to the same conclusion.

This is how the light cones would look like if we don't manipulate the basis coordinate system in the pseudo-universe in the Minkowski spacetime:

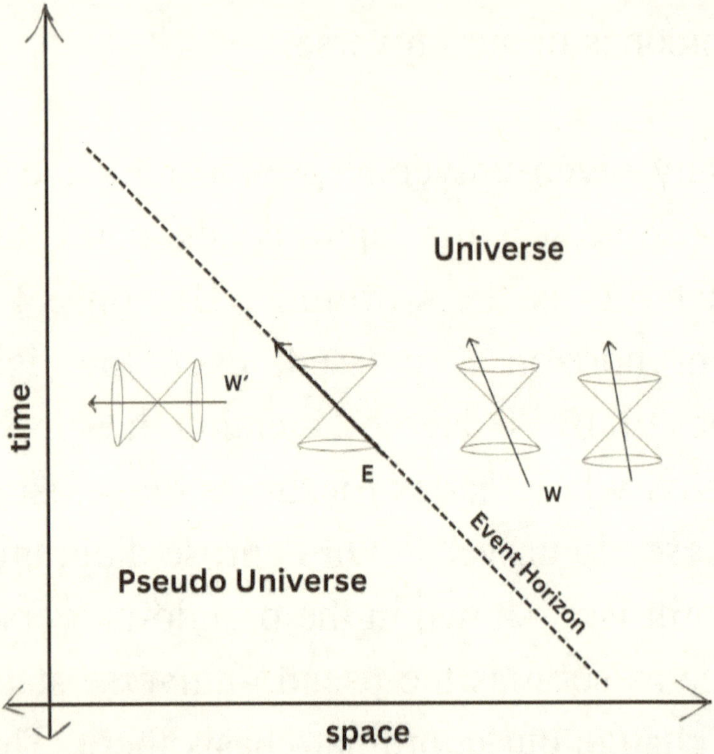

The image shows the tilt of the light cone in the pseudo-universe. The basic coordinates remain the same as our universe. That's why we can clearly show one. In the diagram we are studying the pseudo-universe with respect to the spacetime coordinates of our universe.

The Dotted line in the image represents the event horizon. W is the worldline of a particle before it comes in contact with the event horizon. E is the worldline when the particle is on the event horizon. and W′ when the particle is in the pseudo-universe.

One can see a 90-degree tilt in the light cone as one transits from the universe to pseudo-universe.

As one, can see in the image, when the object crosses the event horizon, the light cone tilts by 90 degrees. This marks the beginning of the pseudo-universe. This tilt in light cone isn't highlighted in Penrose diagrams because we change the basic system boundaries as soon as spacetime switches its roles beyond the event horizon.

In the pseudo-universe, the same rules of special relativity apply. The worldline always has to be within the lightcone.

Forty-three

Less than Zero Speed

Motion is a fundamental thing in the universe. Spacetime supports the existence of matter. Matter defines the spacetime by its asymmetric behavior in one of the four dimensions of the infinite hyperspace. When one speeds up in space, one slows down in time, and when one speeds up in time, one slows down in space. This has been revised multiple times in this book. So, everything existing in this universe is in constant fundamental motion in spacetime. And that we all travel at "c" in spacetime!

The "0" speed and "c" are both the speed barriers in the universe. Nothing can go faster than "c" and slower than 0 speed.

Firstly, we don't know what direction to move to attain slower than 0 speed. And in case of "c" we think that we know, but no direction works out.

Special relativity has its solid boundaries. But general relativity is giving the domain to imitate the worldlines following the spacetime curvature that is strong enough to break the rules of special relativity globally. The rules of special relativity can be broken globally. For example, being over the Event horizon is the same as travelling at "c" in space.

The speed more than "c" or less than 0 speed doesn't come by boosting the booster in any given direction. And also it is not supported by the rules of special relativity. But general relativity has given us the domain to attain this again! As explained in the previous chapter, travelling faster than “c” is same as crossing the event horizon of a black hole. After which, the particle enters the pseudo universe. Starting again from 0 speed in space and “c” in time. Domains like these allowed by the general relativity help us answer the hypothetical questions of special relativity. For example, what happens when one hypothetically travels faster than “c”. At event horizon, a particle with some mass has the worldline as if it’s travelling at “c”. When it enters the pseudo universe, it is that as if it goes faster than “c”. Which is nothing but 0 speed in the pseudo universe. And the less than 0 speed is nothing but “c”.

What we learn from relativity is that the universe changes itself just to keep speed 0 to “c” speed range constant wherever it is within the universe. The 0 speed is the infinitely small speed of the universe, and "c" is the infinite and max speed attainable of the universe.

Forty-four

The Singularity

General relativity is robust. It gives particles the worldline they would never have gotten due to the constraints of special relativity. It doesn't only stop at the pseudo universe, it goes well beyond that. Creating something where the matter doesn't switch the roles, but instead decides not to exist. We are talking about the singularity of a black hole. Look at the image below:

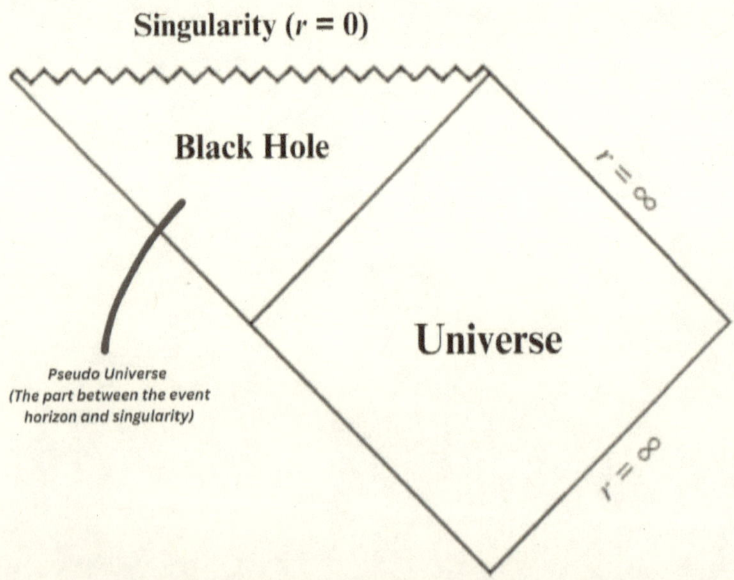

The wiggly line in the image represents the singularity. Wiggly line because it is a breakoff! A breakoff from the universe.

Motion is the fundamental characteristic of matter that marks its existence, and it is what defines the spacetime itself. It is a behavior of matter to speed up in space and slow down in time and speed up in time and slow down in space that defines the existence of spacetime. On other hand, spacetime supports the existence of matter-energy. They both need each other, and they coexist. In the absence of this behavior, the matter doesn't exist, and the spacetime is just one of the dimensions of infinite hyperspace. The singularity marks this point, where matter ceases to exist. It is a point where spacetime and matter both cease to exist. It marks the solid boundary in the pseudo-universe. The boundary of non-existence.

The pseudo-universe is just the universe with a switch in the behavior of matter. Or else everything is fine there as it is in our universe. But as matter moves in time (which is its primary role), it encounters non-existence. Which is another name of Singularity.

When one is under the gravitational influence of a black hole, one physically falls into it until one reaches the event horizon of a black hole. After the event horizon is crossed, the concept of “falling into the singularity” is incorrect. In the pseudo-universe, singularity is not a point in space, it’s a moment in time. The pseudo-universe is a new universe element, and the singularity is a “cutoff” in it.

Hiten Shelar
@HitenShelar

The singularity is a point in space when looked from the outside universe. The singularity is a point in time when looked from inside the black hole. Relativity stuff.

#relativity

22:47 · 26 Mar 25

According to special relativity, nothing can escape the singularity, as it requires faster than "c". But since we know how robust general relativity is, let's know that case.

Forty-five

Escaping the Singularity

Many times the general relativity changes the worldlines of particles in a way that is not permissible in special relativity. The singularity lies in the pseudo-universe part of a black hole. And no matter where one goes in space, one can't escape it. The singularity is not a point in space and not a moment in time, it's both. It's an event. It's looked at as an inevitable event for anything that is in the pseudo-universe part of a black hole.

There is a solution to escape the singularity of a black hole. But one has to be slower than 0 speed for it. If one goes slower othan 0 speed in the pseudo-universe, one can actually return back to our universe. And escape the singularity. Or one can travel faster than "c" in the pseudo-universe and enter another universe. Which is not a pseudo-universe but just like some other universe like us. The parallel universe!

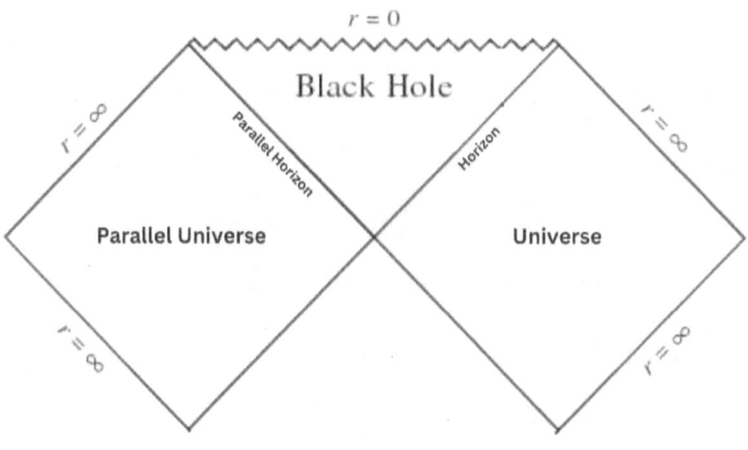

The special relativity constraints exist wherever there is spacetime. Even in the pseudo-universe. But if one can manage to go beyond "c" with general relativity. One can make it to the parallel universe. A black hole in the pseudo is not a good candidate since a black hole creates singularities that won't take us to the parallel universe. But if by some way if we achieve the worldline tilt more than 45 degrees, we can surely make into the parallel universe. So, from our universe to pseudo-universe and pseudo-universe to parallel universe would be the journey in such a case.

The parallel universe here is just a result of a switch in the behavior of matter inside the pseudo-universe. The same switch in behavior that matter chooses to go from the universe to the pseudo-universe. This way one can escape the singularity.

Topologically, the light cones in the parallel universe would be 180 degrees tilted as compared to the light cones in our universe. It can be seen as a reversal of the direction of how matter travels in the dimensions of time. Still, the entropy wouldn't decrease in such a universe (Explained in one of the preceding chapters).

Forty-six

The Slowest and the Fastest Speeds of Our Universe

The speed of causality "c" is the fastest speed of the universe. It is the infinite speed of the universe. At the same time, the 0 speed is the slowest speed possible in our universe. Nothing can travel faster than "c" and slower than 0 in spacetime. The speed faster than "c" is 0 speed, and the speed slower than 0 speed is "c". This is how matter maintains it's existence throughout the universe, pseudo-universe, and the parallel universe.

In whatever speed one travels between 0 to "c". There are two things in common. First, to be at absolute rest in one's own frame of reference. Second, the constancy of the speed of causality. These two things are constant at any frame of reference.

In space, there are an infinite number of directions to speed up a particle toward. The speed that a particle can gain is between 0 to "c". But at these two extremes (0 and c), the concept of an increase and decrease in speed loses its meaning. For example, in what direction will one start moving to move slower than 0 speed? A similar question arises when one attains the speed of "c". For a photon that travels at "c," there is no direction that will help lower the speed. There is no direction to move anything faster than "c" and slower than 0 speed. Going faster than "c" and slower than 0 speed will just result in the entrance into a "new spacetime". Sometimes, the pseudo-universe, and other times the parallel universe.

Nothing in this universe can exceed the speed limit “c”. But in the case of black holes, the spacetime distortion itself results in tilting of more than 45 degrees! This does break the rules of special relativity globally (which is very common globally). Special relativity says that no energy present in the universe summed up can accelerate a particle with non-zero mass to “c” and “faster than c” in “spacetime”. The acceleration or variation in speed manifests as a tilt in a worldline in the spacetime. According to special relativity, this tilt is not possible to exceed the 45-degree mark. But with general relativity, one looks spacetime as a geometry that changes around the total mass-energy density. And in the universe, some sources change the spacetime so drastically that the worldlines of the particles within the distorted spacetime attain and go beyond 45 degrees (Black holes). Even after the particle's worldlines get tilted, the particles continue to obey the rules of special relativity in the pseudo-universe by switching how they behave in spacetime.

The special relativity talks about the existence of matter and its relation with spacetime. It talks about the things that give spacetime its meaning and set it apart from the infinite dimensions of hyperspace. General relativity is more of knowing a fact and that the structure of the universe is a geometry that can be changed by mass-energy density.

Forty-seven

When Gravity Blinks

Like there is a spacetime curvature, there are wiggles in the fabric of spacetime. These waves in the fabric of spacetime are called the gravitational waves. It has the same effect as tidal gravity in anything it passes through. The tidal gravity is basically a difference in gravitational pull in a body due to the difference in proximity of both ends of the body to the source of gravity. For example, the front side of the moon is more near to the Earth. And it goes through more spacetime curvature than its opposite side, making the side facing the Earth forever. This is called the tidal locking.

The gravitational wave has a peak and trough like a normal wave. Anything this wave passes through, goes through a tidal force as trough and peak of a wave pass through the object. Whenever the gravitational wave performs a non-zero work by passing through any object, its energy decreases.

When two or more massive objects like stars, neutron stars, black holes, etc are on the verge of collision, these waves are created. Like light, these waves travel at "c". And they mark their omnipresence in the fabric of spacetime. An it's we who are encountering them at "c" in time.

Gravitational waves are the pulses of gravity. When two light sources are orbiting each other, they eclipse each other and create flashes of light (Pulses of light). Similarly when two neutron stars or black holes revolve around each other before the head on collision, they eclipse each other's gravity. Thus creating the pulses of gravity, the gravitational waves.

The gravitational waves get their energy from nowhere but from the mass of the source itself (mass-energy equivalence). For example, when two neutrons are in a head-on collision with each other, they produce gravitational waves. Now, let's say the gravitational waves pass through a rubber ball and the rubber ball stretches and squeezes accordingly. In the process, tension builds in the rubber material of the ball, increasing its temperature. Astoundingly, this heat energy comes from nowhere but from the mass of the neutron star itself! This way, the gravitational waves literally radiate the mass of the source.

The waves of gravity or gravitational waves aren't something that we have never encountered. It's just a gravity for a limited amount of time. One must not forget that one is already under the continuous encounter with the gravitational waves as one travels at "c" in time. It's nothing new but the spacetime curvature of the Earth. The spacetime curvature due to Earth can be looked as a continuous source of gravitational waves with now trough and peaks. Troughs and peaks only generate when spacetime curvature is eclipsed by another spacetime curvature. Like a constant source of light, the Earth is a constant source of gravitational waves (uneclipsed). When the gravity blinks like a blinking light, the individual waves of gravity are created.

When the spacetime curvature blinks, it produces the gravitational waves

To get an idea of a frequency, one can imagine the gravitational waves due to Earth as the radio waves and the gravitational waves due to colliding neutron stars as the X-rays (Just an example).

The stronger and sharper gravitational waves need two things. First, a huge spacetime curvature like that of the neutron stars and blackholes. Second, a very sharp blink. A blink this sharp can only be produced when two neutron stars or black holes are on the verge of collision with each other and are close enough. At first, two neutron stars come near to each other, and as they get super close, as they revolve around each other at very high speeds along with eclipsing each other's gravitational fields, they get combined into a black hole. The action of eclipsing each other’s gravitational fields produces individual spacetime curvature blinks, aka the gravitational waves.

We have observed a lot of gravitational waves in the universe. None of them are seen to be as sharp to affect a human-sized object significantly.

Due to their sharpness and strength, they ripple across the universe like light. Their energy decreases as they damp into performing some non-zero work. Gravitational astronomy is a new branch of astronomy pushing our limits to study the furthest objects even better.

Forty-eight

The Expansion Of The Universe

Our universe is expanding, it's doing it faster than "c". The further one looks in distance, the more the universe is seen to expand. It's seen to be expanding faster than the speed of causality. The galaxies sitting on it are moving faster than "c". This is not a violation of special relativity because Galaxies are not doing it locally. And so, no time dilation and length contraction are seen for these galaxies. If they had really travelled near "c" in space, the effects of length contraction and time dilation would have been visible on them. No such effect is seen on them. The universe can expand faster than the speed of causality, but not the things living in it.

The expansion of the universe is not like the expansion of the balloon. It doesn't stretch! The universe is expanding because new spacetime is emerging all over the universe. This happens every single smallest unit of time all over the universe. At least this is our observation. The electromagnetic bonds between the atoms of our body and everything we see are far stronger than the repulsions generated due to the expanding universe at our scale, so it's not a threat to anything even as huge as a galaxy. The effects are only visible at cosmic distances.

The expansion of the universe means new spacetime has been created within the universe. But wait! Does that mean new mass-energy? The first law of thermodynamics says that the total amount of energy in the universe always stays constant. Not many people know that this law only holds true locally. And that the energy globally is not constant.

The spacetime supports the existence of matter. And matter gives spacetime its meaning. Matter-energy speeds up in space when slowed down in time and speeds up in time when slowed down in space. This asymmetrical behavior of matter in dimensions of hyperspace makes spacetime existent. Otherwise, spacetime is not different from the hyperspace.

When we say that the spacetime is creating itself, it has to mean that the zero point energy defining the spacetime has to add up in the total energy content of the universe. Since new spacetime requires this basic energy content associated with it. We have discussed that the 0-point energy is the lowest energy possible at any event of the universe (any point in spacetime).

Since the Big Bang, the total energy has been increasing in the universe globally. The new spacetime supports the existence of added energy, and the added energy defines the newly created spacetime. This is the interplay of existence between matter and spacetime. But the question one might ask is what causes the mass-energy to increase at the first place? Let's explore it further.

Forty-nine

Is The Universe Really Expanding At All?

We have pretty good evidence of the expanding universe. The evidence is the redshifting light. The light from the objects at cosmological scales is seen to have a drop in its frequency. The creation of new spacetime at cosmological scales steals the energy of light, making it fall to lower frequencies.

In past chapters, how we interact with the light has been explained. It has been made clear that the universe is a 4-dimensional entity. So, instead of just believing what we

see, applying a higher-dimensional logic is necessary to understand the reality.
Light or photon is an omnipresent thing. It marks its existence throughout the spacetime. So it is we who are interacting with the already present light in spacetime. And we do it at “c”. The redshifting in light is an omnipresent thing,

The omnipresent light shows us the dimension of our universe. The dynamism of the increase in the dimension of our universe in our observation comes from nowhere but from our motion in time at “c”. It tells us that the universe is already expanded in whatever size it is. And we are moving in time in the direction where the universe has an increasing expanse. This doesn’t mean that the universe itself is expanding.

And the faster-than-light-moving galaxies underline the topology of the already expanded (at least at some scale) universe. The galaxies appear moving faster than "c" due to their motion in spacetime that's already in the expanded state.

The rate at which we witness the expansion of the universe is based on our speed in the time dimension. Any type of evolution in the universe that we witness is based on our speed in the time dimension. As we slow down in time, the faster the universe appears to expand. For example, the galaxy moving at 2 times "c" will move at speed of 4 times "c". For anything that travels at "c" in space and so rest in time, the universe is already in its expanded state.

The universe is a four-dimensional structure. Its geometry shows its effect on how matter evolves in spacetime.

Considering that time is four-dimensional—something the human mind can't think of—and our motion in it at "c" . We get to see how things actually work.

Fifty

Omnipresence and Dynamism

If the light is omnipresent in spacetime, does that mean that it was predetermined in the light's perspective for you to see it? These questions might be the ones that might have come through in one's mind throughout. Also, we have talked about the concept of presentism in this book, which is how the universe is, in my opinion.

So, is it predetermined in light's perspective to be observed by the observer? Making us think about whether it is the omnipresence or dynamism.

It might be tough to accept the dual notions at a time. But it will make more sense after the next chapter on “The Permanent Energy JACKET”.

Light shows its omnipresence while being in the universe where it is dynamically created. The light is omnipresent as soon as it's created. But light itself is created in a dynamic universe. Also, it’s the dynamic universe that decides if light ceases to exist or not. If we make contact between an electron and a positron, they will annihilate, releasing an omnipresent photon. But the act of creating that photon was based on dynamism. And it was “our choice” to do it. In case of any other hypothetical particle that travels at “c”, the same applies.

Fifty-one

The Permanent Energy JACKET

The energy is like a jacket that can only be seen when something wears it. And the only way to take this jacket off is by transferring it from one system to the other.

In the past chapters, we discussed semi-permanent energy jackets, temporary energy jackets. The temporary energy jackets are based on the semi-permanent jackets. Something that is more general to the semi-permanent jacket is the "Permanent Energy Jacket". This is the original source. Semi-permanent jackets are semi-permanent because they are just the "spacetime" form of Permanent Energy JACKET. Semi-permanent and temporary Jackets present or not present, all belong to this JACKET. The Permanent energy JACKET. The Permanent Energy JACKET is non-transferable. It can take forms, those forms are transferable. But at the end, everything falls into the category of Permanent Energy JACKET. The forms it takes are like the form it never took. It's the original source. The Permanent Energy JACKET is a Hyperspace thing. It transcends the spacetime. Where creation and non-creation both belong to it.

Just after the Big Bang and before the Higgs field gave particles their respective masses, all the particles were travelling at "c". They had no rest mass. Nothing with zero rest mass has proper time. No time ticking at all! To recall, it's ability of matter to travel in space at "c" in space when it is at rest in time. And travel at "c" in time when it is at rest in space and fractionally both that defines the concept of spacetime. And spacetime supports the existence of matter-energy. They both co-exist. When every particle travelled at "c" in space, there was nothing moving in time. The motion in time started when the particle received its mass. The dimension of time existed, but it was just like a road that no one was traveling on.

As soon as particles got their rest mass, they started traveling at "c" in time, thus giving rise to the concept of "Proper Time". But not all the particles received a rest mass. For example, the photon. The photon still abandons the dimension of time. It treats the universe as a Hyperspace while existing in it.

As we travel in time into the already expanded universe, we see more and more

Permanent Energy JACKETS taking the form of Semi-permanent energy JACKETS that are transferable to temporary energy JACKETS. Thus, energy globally is not constant. Right now, it is increasing.

The semi-permanent and temporary JACKETS were all in their original form of Permanent Energy JACKET before they were dressed up with spacetime as we head our way in the already expanded universe through our motion in the dimension of time. The Permanent Energy JACKET is a Hyperspace thing. It is, in fact, omnipresent. In this book, the term omni presence means spread all over both in space and time. There is no gap of "time" between a cause and effect. The cause and effect both exist in the same frame. In fact, there is no such thing as cause and effect. Every state of evolution exists in one frame. This is how the world of a "photon" works. The photon or light is kind of part of both, the universe and the Hyperspace. Thus, the massless (frequency) and mass are given the term "Semi-permanent" Energy JACKET. The Permanent Energy JACKET is a Hyperspace thing.

The creation and absorption of a photon is rooted in the cause and effect game of the universe, but not so its existence! Its existence transcends the dimension of time. And when you transcend the dimension of time, you become a Hyperspace entity. This applies to every particle that travels at "c" in the space of spacetime.

As we move at "c" in time into the already expanding universe, the Permanent Energy JACKET is seen in the form of what we address as "energy" in spacetime. The Permanent Energy JACKET is the original source of all. It's a Hyperspace thing.

The universe is just another creation of Hyperspace. In a general sense the universe as a whole itself is omnipresent in reality. Everything belongs to the Permanent Energy JACKET. The Permanent Energy JACKET is a Hyperspace thing.

Fifty-two

The Omnipresent Energy

The energy is the ability of a system to perform some work. The stretching of the omnipresent light tells us that the universe has already expanded in its current state. And it's our motion in time that makes it appear as if it is dynamically expanding. But the already expanded spacetime universe needs energy to define its existence.

The universe can be seen as a cave, whose mouth width keeps on increasing in one direction as one moves. The direction represents the direction of our motion in time at "c". The increasing mouth width of the cave is what we perceive as the expansion n of the universe at any given slice in spacetime. This slice right now is increasing in size as one travels at “c” in time. This slice supports the existence of matter.

But for this slice to be defined in the first place, it requires a non-zero energy value to be present underlining its existence. Not only for this slice, this is true for the entire already expanded current state universe. Does that mean that energy exists in the future? Correct! But not exactly as one most probably might imagine.

In physics, energy is like a JACKET that can never be destroyed but only be transferred from one form to another. We see the energy in the form of matter carrying it. And we have never defined this quantity as it’s naked or not carried by matter. Where matter itself is nothing but energy.

The pure energy which I am talking about defines the most fundamental existence of everything in the universe. It marks the starting point of all relativity. When this energy itself diverts from this starting point, it becomes the mass. This pure energy is omnipresent! It can only be seen in the standard of physics when the traveler meets it while traveling at "c" in time in the fabric of spacetime.

Right now, we are heading towards the part of the cave where the mouth width is increasing. And thus we meet the already existing pure energy at the same rate. This appears as if the total energy content is adding to the universe. But it's instead it's we who are meeting the already existing pure energy as move in time at "c". When we meet this energy, we see it in the form of something carrying it. Like in the form of temperature, in the form of mass, etc.

If pure energy is omnipresent, does it experience time? The concept of time arises when matter shows its asymmetrical behavior in time as compared to space, provided that time is just a dimension as space. So the answer is no! The pure energy doesn't show the asymmetric behavior in the fourth dimension and make it time. It is something that joins the real (the universe) with something more real, higher, and general. The topic of the pure energy can take another book to describe.

When we say that our universe is 13.8 billion years old, that means that matter started behaving asymmetrically in one of the 4 dimensions of hyperspace. Making that dimension "time" around 13.8 billion years ago. Before which was nothing but the pure energy (formless) discussed in this chapter.

For this pure, formless energy, there exists no time. It's the 7 behavior of matter to travel at "c" in time while resting in space and rest in time while travelling at "c" in space that creates the spacetime.

Fifty-three

Relativity to Reality

The pure energy is formless and there exists no time for it. It's the asymmetric behavior of matter to travel at "c" in time while resting in space and rest in time while travelling at "c" in space that creates the spacetime.

Relativity begins when time comes into the play. The asymmetric behavior of matter gets "time" into the play. The pure formless energy gets a form when matter comes into the role. And the matter is nothing but the energy, the same formless energy!

When the formless energy gets a form, it introduces the concept of time by becoming the matter. When the concept of time is introduced, the theory of relativity arises.

There is no relativity without time. Energy is "formless" without time. And it's the same formless energy that behaves asymmetrically to define time. Giving rise to the material world. Which in our case is 13.8 billion years old.

Thus, the birth of the universe is the birth of time, and the birth of time is the birth of the topic of relativity. Creating the universe as it is! The Universe Is Just a Relativity In Hyperspace.

More Books By Author:

1) Astrophysics For Non-mathematicians.

2) Cosmology For A Newbie.

3) Astrophysics For Everyone.

4) Astrophysics In The Form Of a Story.

5) Astrophysics and Cosmology For Non-mathematicians.

6) Why Are We Alone In This Universe: Destroying The Delusion Of Science Media.

7) The Next Level Observer Effect: Taking Observer Effect to the Deeper Level Of Application.